Three Mile Island: Prologue or Epilogue?

Three Mile Island: Prologue or Epilogue?

by
Daniel Martin
University of Baltimore

Ballinger Publishing Company ● Cambridge, Massachusetts
A Subsidiary of Harper & Row, Publishers, Inc.

International Standard Book Number: 0-88410-629-6

Library of Congress Catalog Card Number: 80-11067

Printed in the United States of America

Library of Congress Cataloging in Publication Data

Martin, Daniel, 1948-
  Three Mile Island.

  Includes bibliographical references and index.
  1. Three Mile Island Nuclear Power Plant, Pa. 2. Atomic power-plants—
Pennsylvania—Accidents. I. Title.
TK1345.H37M37   363.3'497   80-11067
ISBN 0-88410-629-6

Table of Contents

List of Figures

Acknowledgments

A debt of gratitude is appropriate from any researcher on this topic to the staff of the President's Commission on the Accident at Three Mile Island. Using the power of subpoena, they have generated an immense supply of written information on the accident that would not have been available through traditional research channels. In a similar but less complete fashion, the Pennsylvania Select Committee on Three Mile Island has provided information on the activities of local officials during the crisis.

On a more personal level, I want to thank four people. Dr. Neil Kleinman provided valuable assistance in the process of preparing the book for publication. Rick Young took time from his busy days to turn my scratchy drawings into the diagrams in Chapter 2. Ralph McCoy was very helpful in introducing me to the behavior of valves and pumps under conditions of stress. Finally, I want to thank my wife Janat, who once retired forever from secretarial duties, but who volunteered to come out of retirement to help speed this project along.

Three Mile Island:
Prologue or Epilogue?

Chapter 1

A Reactor in Search of a Home

In the chilly early morning mist of March 28, 1979, a small power boat weaved its way along its regular course in the Susquehanna River just south of Harrisburg, Pennsylvania. Three members of the Ichthyological Associates of Goldsboro were on board. Occasionally, they stopped the boat and drew water samples to be taken to the lab for later analysis.

The Susquehanna in the area of Goldsboro is peaceful at dawn. Goldsboro is the only township in sight, and even there only a few homes and one marina line the bank of the river. On the other bank, the township of Middletown is a couple of miles up the road. Goldsboro is a good fishing area, with the water running calmly and relatively deeply compared to the often rocky course of the Susquehanna. But with the distrust in modern society, the local merchants occasionally check to be sure that the townships and the city of Harrisburg upstream are keeping the river clean.

The motor boat broke the quiet as the three men moved past the tip of Brunner Island toward the only source of industrial pollution. They moored the boat at the next island and climbed out to draw a few water samples along the shore. At their backs, less than one hundred feet away, was a fence. Just beyond that was another higher security fence. Beyond that were the enormous cooling towers and the rest of the imposing complex of the nuclear powerplant on Three Mile Island.

None of the people in the boat were strangers to the powerplant. It had been there for years and had been a reasonably good neighbor. During the day, it was a hubbub of activity. But before the 7:00 A.M.

1

shift arrived each morning, it looked almost abandoned. So long as the intruders kept their distance, the only sign of activity inside was the heavy mist escaping hundreds of feet above from the top of the cooling towers. Even that was quiet.

The ichthyologists collected their samples and returned to the boat. They cranked their engine and continued on their normal rounds. They left behind the quiet buildings, the unwelcoming fence, the rising mist, and what still less than one hundred people knew to be the most serious crisis in the history of commercial nuclear power. Within those buildings, the future of nuclear energy was deteriorating.

This is an attempt to tell the story on the other side of that fence. It is a recollection of the accident from the accounts of the people in the plant, in the state emergency preparedness machinery, and in the relevant federal agencies. It is not an attempt to condemn or indict any of the actors, although it is not always reserved in tone. Individual performances are attacked only when they exceeded the bounds of reasonable expectations, as they occasionally did.

To judge the accident at Three Mile Island is too close to judging nuclear energy's future role in American society. The precise connection of the two is a decision that must be made by each individual. But to make the connection, the accident at Three Mile Island needs to be reconstructed. That is the goal of this book.

Probably the largest irony in the role Three Mile Island's Unit II is playing in the fate of nuclear energy is that no one in the industry really wanted to put a reactor there. It is located in central Pennsylvania where coal is plentiful and population is sparse. The state capital, Harrisburg, is just twelve miles away, but it is served by a different utility company. Metropolitan Edison, the company that operates Three Mile Island, already had a nuclear powerplant under construction on the island in the 1960s. It was supplemented by coal-fired plants using coal from the western part of the state. To build more of such plants in the late 1960s seemed folly and was not considered.

But the area around New York City was a different matter, and it is there that the story of Three Mile Island's Unit II began. During the 1960s, the suburban sprawl around New York City crept ever further into several surrounding states. Consolidated Edison once proposed building a nuclear powerplant in the borough of Queens, but the Atomic Energy Commission vetoed that request because of the absurdity of evacuation planning at such a site. Instead, the reactors moved to the suburbs with the expanding population.

Reactors and planned reactors began to show up in Connecticut and Westchester County, New York. In New Jersey, plants were started in Salem and Oyster Creek. At Oyster Creek, the unusual history of Three Mile Island's Unit II began to develop.

Oyster Creek's nuclear powerplant is operated by the Jersey Central Power and Light Company and provides electricity to much of the central part of New Jersey. In 1968 it was in the final stages of construction. While it was not yet operating, the population of this once rural area was expanding at such a rate that the designs for a second nuclear powerplant at the same site were already well underway.

Planning and financing a nuclear powerplant is an impressive and risky achievement for a traditional electric power company. There are decisions to be made that require a type of expertise that a utility company would have no reason to develop. The capital and credit arrangements are immense compared to the normal operating problems of such a company, and mistakes in the early phases can cost dearly later.

However, Jersey Central Power and Light was not tackling this project alone. Since 1946 the company had been owned by a larger holding company, the General Public Utilities Corporation or GPU. GPU owned three such power companies serving about half of the population of Pennsylvania and New Jersey. In addition to Jersey Central, GPU owned the Pennsylvania Electric Company, operating out of western Pennsylvania. GPU also owned a small company in central Pennsylvania, Metropolitan Edison.

Normally, GPU maintains control over the subsidiary companies by naming the head of its board and chief executive officer to be the head of the board and chief executive officer of each of the subsidiary companies.[1] But GPU chooses not to get involved in the day-to-day problems of managing a utility company. For that, each subsidiary company maintains a president, and operating problems are handled at the local level.

Until 1967, nuclear powerplants were a local responsibility with substantial assistance from GPU. But with both Metropolitan Edison and Jersey Central building plants, there were obvious advantages to sharing resources and coordinating activities at the top. Therefore, in 1967 GPU formed the Nuclear Power Activities Group as another subsidiary of its corporate headquarters. Once the group was formed, while each of the three operating utilities was responsible for running its completed plants, the group coordinated the planning and the building of the plants from GPU headquarters.[2]

At Three Mile Island, the change of leadership presented no problems. Construction was proceeding slowly, but reasonably smoothly. But by 1968 the situation at Oyster Creek was deteriorating rapidly. Unit I was almost completed, and the design phase for Unit II was substantially finished. On the surface progress seemed impressive. However, an amazing new snag developed in the plans for Unit II in the latter half of 1968.

Oyster Creek, New Jersey, is close to New York, and is similar to New York in some of its business and labor practices. Neither labor nor management in the United States has a consistent history of integrity in its operations. In most areas, corruption takes the form of individualized acts that are easily absorbed by the rest of the community. But when things turn sour in the New York–New Jersey area, they often get noticeably worse than in some other parts of the country. In 1968, things turned sour.

The corporate heads at GPU have almost all changed since 1968, and the new leaders are hesitant to describe a period in which they did not personally participate. Nevertheless, there are enough depositions and legal convictions to state with some certainty that the unions working on the Oyster Creek construction site in 1968 fell or pushed themselves into the grasp of New York's organized crime community.[3]

To GPU, this was manifested when a union local president approached the corporate leadership in late 1968 with an extortion demand.[4] He offered GPU industrial peace for the duration of the construction of Unit II in exchange for a personal donation of 1 percent of the construction price.[5] If GPU did not provide him with these funds, the unions would prevent the completion of the project.

It was too late to simply scrap the reactor. About 75 percent of the design work had been completed.[6] Besides, much of the hardware for the containment building had already been ordered.[7] Beyond this, GPU was convinced that the population of the service area was expanding in such a way that the extra generating capacity was needed.

The corporate leadership approached the attorney general's office for the state of New Jersey. Details of these contacts are so vague that speculation serves no real purpose. However, whatever was said, GPU President Kuhns came away with the distinct impression that the state of New Jersey was not willing to help GPU fight corruption in the labor movement.[8]

GPU was left with three options. They could accept the extortion demand. It amounted to several million dollars, but that was a loss that would quickly be regained by the promise of industrial peace.

There is no evidence that GPU ever considered this option. If they did, they were painfully aware that such dealings with the people approaching them were reputedly very difficult to break off later.[9] The only way to avoid a lasting and even growing complicity was to spurn the initial offer.

As a second option, they could abandon the project entirely. Such a proposal was considered. However, the sunk costs were in the range of $10 to $20 million.[10] These expenditures would be completely lost. In addition, there would be future energy replacement costs since the extra generating capacity would eventually be needed.

As a third option, they could move the plant away from this labor environment. It would have to be all the way out of New Jersey since the corruption problem extended at least that far. That would also remove the plant from the jurisdiction of Jersey Central Power and Light. However, since 1967 GPU's Nuclear Power Activities Group had taken responsibility for the planning and building phases of reactors. Because of this arrangement, the group could easily move the plant beyond the jurisdiction of any subsidiary company.

As Director of the Nuclear Power Activities Group in 1968, Louis Roddis was asked to study the feasibility of moving the plant and to develop a list of possible sites. The new site needed to be within a reasonable distance of New Jersey, but in a different labor environment. It was preferable if it could be attached to an existing reactor since site acquisition and the required geological surveys would not entail additional costs.

Very quickly, the group limited its consideration to the only obvious choice, which was to add a second unit to Three Mile Island. In November 1968 the group submitted its report on that possibility to GPU management.[11] In that report, the group argued that the only substantial disadvantage to moving to Three Mile Island was a time delay for completion of the project. However, construction labor and operating labor costs were strongly in favor of Three Mile Island. Governmental and public relations were not a significant problem at either site.

The report emphasized, however, that the new unit would have to be built at Three Mile Island using a "minimum change" policy from the plans already approved for Oyster Creek. There was a disadvantage to this decision since the group considered the design engineering firm used to plan Three Mile Island's Unit I to be a better company than the one designing Unit II.[12] Nevertheless, switching design engineers at this phase was considered to be out of the question for economic and timing reasons.[13]

For approximately one month, senior management studied

Roddis' report while deciding what to do. Finally, GPU President William Kuhns called a meeting for December 23, 1968, in the corporate headquarters in New York. With all the affected subsidiary companies and contractors present, Kuhns announced that the Oyster Creek Unit II was being deferred in favor of a second unit at Three Mile Island. To perpetuate the complicated internal structure of GPU, Metropolitan Edison would own half of the new unit. Each of the other two operating subsidiaries would own a quarter. The unit was projected to open in 1973.

To increase the irony of the story, the particular labor leaders who caused the crisis at Oyster Creek were eventually convicted and imprisoned in another extortion attempt. When the labor environment in New Jersey cleared, GPU made a second attempt to put a reactor in the Oyster Creek area. This one was not literally at Oyster Creek, but was within eyesight at Forked River. It was partially finished when the accident at Three Mile Island Unit II caused the project to be halted because of the sudden financial uncertainty the accident imposed on GPU. If there is an external force interfering in our daily lives, it seems determined not to have a second plant near Oyster Creek.

In 1968, Metropolitan Edison found itself to be the primary owner of a second reactor that it had played no role in designing. Despite the lack of intentional coordination, however, the planned Unit II was much more like Unit I than like Oyster Creek. The nuclear steam systems for both were designed by Babcock and Wilcox, whereas Oyster Creek's system had been designed by General Electric. Both units at Three Mile Island were actually built by United Engineers and Constructors.

But there were also substantial differences between the plants that were to become a factor during the accident. These centered around the nonnuclear design engineers. Met Ed used the highly respected firm of Gilbert Associates for Unit I. For Unit II, the company of Burns and Roe was retained, since it was already well into the design for Jersey Central. Unfortunately, as stated earlier, the director of GPU's Nuclear Power Activities Group considered that "Gilbert is a better design engineer."[14]

The nonnuclear design engineer fills an unusual role for utilities building nuclear reactors. The exact parameters of their job are not specified. In fact, Burns and Roe worked on this reactor in both locations from 1967 until 1975 before they finally had a formal contract with GPU. Instead, in the interim, once the nuclear steam components and the turbines and generators were purchased, Burns and Roe's job for Met Ed was to "perform as a service to the client anything the client requires and cannot or does not choose to do themselves. . . ."[15]

In defense of Burns and Roe against the poor impression they made on GPU, they had a good history with Jersey Central, and most of the nonnuclear components were completed to Met Ed's satisfaction. However, there were notable exceptions. For instance, Met Ed was unable to convince Burns and Roe or even GPU that Met Ed should be able to order changes that they felt improved the design. By far the most commonly raised example is the layout of the control room.

In 1967, when the Oyster Creek Unit II control room design was first considered, there was no standard model for control rooms in the industry.[16] Instead, Burns and Roe engineer Edward Gahan visited Oyster Creek I to get a feel for what they could suggest in the way of improvements.[17] Using Jersey Central's recommendations, he designed two optional control rooms to give the utility a choice. Their choice was accepted by Burns and Roe.

But when Met Ed took responsibility for the proposed plant in late 1968, they set about reviewing the existing plans. In early 1969 Met Ed requested that the Unit II control room be completely redesigned to make it similar to the control room in Unit I.[18] This is a common practice in the industry, and several reactors run two nuclear cores from one large control room that is divided into two almost identical halves.

To resolve this request, GPU called a meeting in March 1969 with representatives from Met Ed, Burns and Roe, and Jersey Central, since that company had made several suggestions incorporated into the original room. Burns and Roe and GPU both argued that almost identical control rooms were more confusing to operators than completely different rooms. Instead, any future changes in the control room, including those proposed by the operating company, Met Ed, would not be accepted without the approval of either GPU *or* Jersey Central.[19] Clearly, Met Ed had lost its power in demanding changes in the reactor that it would eventually inherit.

There were other difficulties with the design as far as Met Ed was concerned. The condensate-polisher system, explained in Chapters 2 and 3, was transferred directly from the Oyster Creek plans even though it was designed for salt water there and for fresh water at Three Mile Island. When Met Ed reported on a malfunction in that system during testing in 1977, a malfunction almost identical to the one that began the eventual accident, GPU decided that it was not important and did not forward the report to Burns and Roe.[20]

A chance for revenge did not arise for several years. However, when Unit II reached a chain reaction on March 28, 1978, substantial engineering powers shifted from GPU to Met Ed. One of the first issues to be discussed at Met Ed was whether Burns and Roe should be replaced by a company more to Met Ed's liking.[21] There were several considerations working against Burns and Roe, including the problem that the companies had been working together for years without ever developing a close relationship. While Jersey Central had a friendly relationship with Burns and Roe, Met Ed had developed such a relationship with Gilbert Associates.[22]

However, Burns and Roe had taken the plant from a verbal concept in 1967 to an operating reactor in 1978. In many ways, because of their cooperation with GPU, they knew several aspects of the plant better than Met Ed. As a result, while the working relationship was not overly close, Met Ed entered commercial service with a continuing service contract from Burns and Roe.

On February 8, 1978, Met Ed received an operating license for Unit II from the U.S. Nuclear Regulatory Commission. While Jersey Central and Pennsylvania Electric were named on the license as joint owners, GPU specifically was not. Therefore, while the lengthy testing of nonnuclear systems had been in progress for some time and would continue, only Met Ed employees were permitted to turn knobs in the control room. While the plant was almost three months from generating electricity, it was more clearly becoming the responsibility of Met Ed.

The next step along the road to commercial operation came at 4:12 A.M. on March 28, 1978. The nuclear fuel had already been loaded into the core, but was restrained by the control rods. At 4:12 A.M., the control rods were eased partially out of the core and a chain reaction started. The plant was not yet generating electricity because it was still being tested. However, ten years after the plant was moved to Three Mile Island, and five years behind schedule, the reactor finally had a reaction.

That lasted one day. On March 29, with the reactor still operating at partial power, there was an accident freakishly similar to the one that would occur one day short of one year later. The reactor automatically shut off, and the pressure relief valve (explained in the next two chapters) opened on command and then stuck open. The primary coolant, pressurized to 2155 pounds per square inch (psi), blew out through the open hole until the pressure was so low that the Emergency Core Cooling System engaged. But this was under test conditions, and control of the reactor was quickly regained.

The malfunctions were repaired, and the chain reaction was started again on April 8. This time it ran under test conditions for ten days before a malfunction caused the reactor to automatically shut down. This malfunction was easily diagnosed, and the reactor was restarted the same day. It ran one day and stopped again. It was restarted and ran one more day and then stopped again.

When restarted, it ran for three days. But then it once again had an accident similar to the one on March 28, 1979. Once again, the pressure relief valve stuck open. Once again, the emergency core cooling automatically engaged. The reactor was still under test conditions and was easy to control. However, the malfunctions this time were more difficult to repair than last time, and the reactor was not restarted for five months.

On September 17, it was finally restarted. It ran three days and then shut itself down again. This time, however, when it was restarted, it stayed on for less than one hour.

From September 21, 1978 to the end of the year, the reactor either shut itself off or was manually stopped eleven more times. During October it briefly reached 90 percent of full power, and Met Ed actually connected the electric generator into the power grid. It did not stay that way for long.

Within reactors, there is a bookkeeping term called "going commercial." This term has generated some controversy in the case of Three Mile Island and is generally misunderstood by the public. "Going commercial" has no particular relationship to generating or selling electricity. Unit II first generated electricity on April 21 and first plugged in into their power grid in October.[23]

There are, however, monetary consequences when a utility declares itself to be in the status called "commercial." Most immediately, the utility can calculate the net worth and the depreciation of the reactor into its rate base when it goes commercial. At least, this was Met Ed's understanding. However, later in 1979, several utilities attempted to get the construction costs included before the reactor was even half built.

It was also commonly assumed that Met Ed could gain $55.1 million in federal taxes by declaring itself to be commercial before the end of 1978. However, financial consultants for the President's Commission were later uncertain that this qualified Met Ed for the tax breaks.[24] At any rate, Met Ed apparently believed this to be the case. Finally, and certainly not least importantly, once Met Ed declared the plant commercial, GPU lost what few powers it still had over the operation of the plant.

For a number of reasons just stated, Met Ed raised its core power on the evening of December 30 from 80 percent to almost 100 percent of its maximum theoretical power. It was generating almost 900 megawatts of electricity being fed into the power grid going primarily to northeast Pennsylvania and New Jersey.

At 11:00 P.M. on December 30, officials at Met Ed signed the requisite papers. Three Mile Island Unit II had become the newest member of the nuclear community.

NOTES

1. Deposition of Robert Arnold for President's Commission on the Accident at Three Mile Island, Harrisburg, Pennsylvania, August 11, 1979, pp. 22–23.

2. Ibid., p. 24. Deposition of Louis Roddis for President's Commission, New York, New York, August 27, 1979, pp. 9–17.

3. Deposition of Herman Dieckamp before President's Commission, Washington, D.C., August 15, 1979, pp. 102–103.

4. Deposition of Louis Roddis, pp. 81–82.

5. Deposition of James Neely before President's Commission, Washington, D.C., August 23, 1979, p. 106.

6. Deposition of Louis Roddis, p. 87.

7. Ibid., pp. 87–88. Deposition of James Neely, p. 68.

8. Deposition of James Neely, p. 106.

9. Deposition of Herman Dieckamp, p. 102.

10. Deposition of Louis Roddis, p. 88. Deposition of James Neely, p. 68.

11. Contents reviewed in "Report of the Office of Chief Counsel on the Role of the Managing Utility and its Suppliers" to President's Commission on Three Mile Island, October 1979, p. 32.

12. Deposition of Louis Roddis, p. 33.

13. Deposition of James Neely, p. 72. Deposition of Louis Roddis, p. 90.

14. Deposition of Louis Roddis, p. 96.

15. Deposition of Tom Hendrickson before President's Commission, New York, New York, August 1, 1979, p. 51.

16. Deposition of Salvatore Gotilla before President's Commission, New York, New York, August 2, 1979, pp. 21–22.

17. Deposition of Edward Gahan before President's Commission, New York, New York, August 6, 1979, p. 19.

18. Deposition of Salvatore Gotilla, p. 56.

19. Ibid.

20. Deposition of Robert Arnold, p. 72.

21. Deposition of Richard Klingaman to President's Commission, Harrisburg, Pennsylvania, August 3, 1979, pp. 27–36.

22. Ibid., p. 132.

23. "Report of the Office of Chief Counsel on the Role of the Managing Utility and its Suppliers," p. 62.

24. Ibid.

Chapter 2

The Structure of the TMI Reactor

One of the problems that the Metropolitan Edison Company encountered repeatedly during the Three Mile Island crisis was that the public demonstrated little knowledge of the mechanics of nuclear energy and often avoided learning about it until there was a crisis.[1] This may or may not be an accurate assessment of the population as a whole. However, there is clearly a mystique among a substantial portion of the citizenry that the things that happen in a reactor are far too complicated for anyone except an engineer to understand.

To some degree, this is a valid observation. Schematic diagrams of the circuitry of nuclear facilities are amazing jumbles of engineering symbols and jargon. The actual facilities are so complex and so different in structure that the Nuclear Regulatory Commission (NRC) licenses reactor operators only for specific control rooms. Permission to operate in both control rooms at Three Mile Island, for instance, requires two separate NRC licensing examinations.

However, as engineers and contractors will quickly attest, the operators know these wiring diagrams only about as well as an informed driver knows the wiring diagram of his or her car. Operators do not build plants; they operate them. They do not need to know how to wire them; they need to know how to operate them and to understand them well enough to bypass components when they fail. To gain this knowledge, they have found it useful to categorize the plant into the major systems necessary to produce energy. Then they learn the component diagrams and controls to operate and manipulate these systems.

On a limited scale, this chapter has the same goal. Obviously, it is not feasible to explain reactors in anything near the degree of sophistication required of control room operators. However, it is possible to describe the major reactor systems and their roles in the plant. The description can then be used in the chronology of events at Three Mile Island presented in the next chapter.

In most respects, a nuclear power plant is similar to an electric generating plant burning coal or oil. In fact, many operators receive their early training in coal- or oil-burning facilities. In all three types of plants, a heat source is used to turn water into steam. This steam then turns turbines that rotate the armatures in electric generators to create electricity.

The major difference in nuclear plants is the heat source. Coal and oil plants are heated by burning fossil fuel. Nuclear facilities use heat released during the fission of atomic nuclei.

The nuclei of atoms are composed of protons and neutrons, and the identity of the element is determined by the number of protons in the nucleus. However, the number of neutrons in the nucleus can vary in most of the heavier elements, creating varieties or "isotopes" of the element. The working principle behind nuclear fission is that several isotopes in the heavier elements are unstable. Within a specified period of time, each unstable nucleus will split or "fission," releasing two or more neutrons and a tremendous amount of heat.

This heat can be carried away by a coolant and used for the generation of electricity. However, if one of these loose neutrons is now captured by another nucleus, it too can become unstable and fission, creating more heat and more neutrons. If this sequence occurs continually, it is called a chain reaction.

Of course, most of the neutrons do not hit another nucleus, but fly off and are lost. For a chain reaction to occur, the fissionable material must be so tightly compacted and so large in volume that a fair percentage of the neutrons hit another fissionable nucleus. At such a point, a critical mass is achieved.

There are a number of isotopes of various elements that can undergo fission, but each one reacts differently and requires different environmental conditions for a chain reaction to occur. U-238 for example, the isotope that comprises over 99 percent of natural uranium, gives off relatively few neutrons and absorbs them only if they are moving very fast. This is not an insurmountable problem. However, water, which is used to carry away the heat from the reactor, absorbs some neutrons and slows down the rest. Therefore, a chain reaction with U-238 and regular water (called "light water" to distinguish it from a different isotope) is impossible.

But 0.7 percent of natural uranium is the isotope U-235, which fissions easily when hit by slowed neutrons. Since water slows the neutrons to an acceptable energy level for U-235, it is easy to establish a chain reaction using U-235 and regular "light water." The water acts as the coolant and is also called the "moderator" since it influences the speed of the neutrons. In fact, the reaction is possible when natural uranium is "enriched" from 0.7 percent U-235 to just 3 or 4 percent U-235. This is the fuel and the coolant that are used in light water reactors, which include almost all commercial reactors in the United States.

It should be mentioned that it is not possible for a light water reactor to explode like an atomic bomb. The critical mass required in a bomb is actually "supercritical" and can be achieved only when extremely enriched uranium is crushed together by conventional explosives. Even if reactor fuel were to be blown together—and a reactor has no mechanism for doing this—it would not be enriched nearly enough to explode.

However, criticality can be lost over time in a reactor. Eventually, the fuel fissions away so much of its U-235 that there is not enough left to sustain a reaction. As a result, reactor fuel is replaced about every three years.

A similar principle is used to shut off a chain reaction when that is desired. Boron or cadmium control rods can be inserted among the fuel elements, and they absorb so many neutrons that not enough are left to carry on a chain reaction. This can also be accomplished by dumping large quantities of boron solution into the cooling water. When the boron or cadmium is removed, the chain reaction begins again.

Three Mile Island is a light water reactor. It uses uranium enriched to about 4 percent for fuel, boron control rods, and light water as a moderator and coolant. The fuel is in the form of uranium dioxide, formed into small cylindrical pellets less than an inch long. The pellets have a zirconium alloy coating to seal in the uranium and decaying elements that become radioactive during use. These pellets are loaded into zirconium alloy hollow rods about twelve to thirteen feet long. These rods are then connected into bundles of almost 200 to make a fuel assembly. The fuel assembly is about ten inches square and twelve to thirteen feet long. The TMI core is comprised of 177 fuel assemblies, containing about ninety tons of uranium dioxide. It also contains slots for well in excess of one hundred control rods.

Beyond these elements, there is one more major variation in light water reactors. One type is the boiling water reactor (BWR), produced in the United States by General Electric. In this design, the moderating water is heated by the core to about 550° F and about 1000 pounds per square inch (psi) pressure. The water boils to steam, and the steam drives the turbines. The steam is then cooled and condensed back to water to prevent overheating the system and to allow for a temperature differential to encourage circulation. This water is then pumped back to the core to be reheated. The design of a boiling water reactor is shown in Figure 1.

The second major design for light water reactors is the pressurized water reactor (PWR), produced in the United States by Babcock and Wilcox, Combustion Engineering, and Westinghouse. This design gained a competitive edge early in the nuclear industry because of its extensive use and testing in the nuclear submarine program, and it now comprises over 60 percent of the reactors in commercial operation in the United States.

The philosophy behind the PWR is that it attempts to minimize the flow path of core-cooling water in the plant. Instead of flowing through the turbines, the water leaving the core in a PWR goes directly into a steam generator. Here it transfers heat but almost no radioactivity to a second closed water system that flows to the turbines. The design is shown in Figure 2.

This design has the advantage that the somewhat radioactive cooling water is confined to a small loop that can be entirely housed within a containment building. Unlike BWRs, where components such as the turbines are in moderately radioactive and highly restricted areas, most of the components of a PWR are easily accessible for emergency and regular maintenance. As a trade-off, however, PWRs like Three Mile Island have several more sophisticated components that can and do develop problems.

One of the more obvious components in pressurized water reactors that is not in boiling water reactors is the pressurizer. Unlike boiling water reactors, the cooling water in the core of Three Mile Island and other PWRs does not boil. Only the secondary cooling water converts to steam, as needed to drive the turbines. Even though the primary water reaches normal operating temperatures in the neighborhood of 580° F, it is pressurized in excess of 2000 psi to prevent it from boiling. This is done because hot water transfers heat in the steam generator much better than steam can.

However, it is essential to maintain adequate pressure in the primary coolant loop, or the very hot water will flash into pockets of

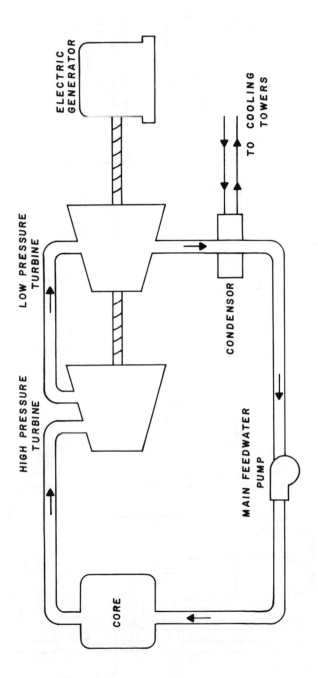

Figure 1.   Boiling Water Reactors.

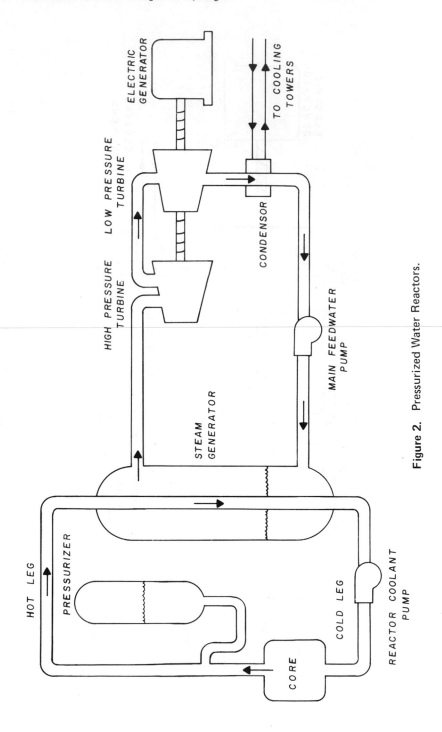

Figure 2. Pressurized Water Reactors.

steam called "voids." These voids can destroy pumps designed for water. The voids can also congregate in the elevated parts of the coolant loop and can eventually block the flow of water through the cooling system.

There are counteracting measures available to fight voids should they form. However, life in the control room is much easier if the voids are prevented, even during temperature and pressure fluctuations. The tool for doing this is the pressurizer, a large container connected to the cooling lines between the core vessel outlet and the steam generator. A diagram of the pressurizer is shown in Figure 3.

The job of the pressurizer is to adjust the water pressure in the primary system to maintain the amount needed to prevent voids in the cooling system. Normally, this is a simple process. However, if the reaction in the core is shut off or "scrammed," the temperature and pressure drop dramatically, and the volume of water in the cooling system "shrinks." To counteract this, the pressurizer controls add water to the system through up to three "makeup" pumps. More pumps are available elsewhere in the primary system should there be a more serious loss of coolant accident.

If the core-cooling water should overheat, as it does if the secondary cooling system is lost as a heat sink, the temperature and pressure climb. The primary coolant loop can quickly overpressurize and damage the equipment or theoretically even explode. In such a case, one of several valves opens on top of the pressurizer to release the excess pressure.

All PWRs have code safety valves that open to release large amounts of water during serious overpressurizations. The excess water and steam are diverted to a holding tank, often called a quench tank. Most PWRs, including Three Mile Island, also have a pressure relief valve set to open at lesser overpressurizations. At Three Mile Island, this is sometimes called the pressure relief valve, the pilot-operated relief valve, or the electromatic relief valve. To minimize confusion, it is called the pressure relief valve here. When this valve exists, as at Three Mile Island, it opens at low overpressures and is often intended to be used in less than emergency conditions.

While the purpose of the pressurizer is to control pressure, the mechanism for doing this is to control the volume of water in the primary coolant loop. For any temperature, as determined by the heat given off in the core, there is a certain volume of water appropriate to maintain adequate pressure. Adequate pressure is the amount necessary to avoid the formation of steam voids in the primary coolant loop. Ironically, the way the pressurizer monitors

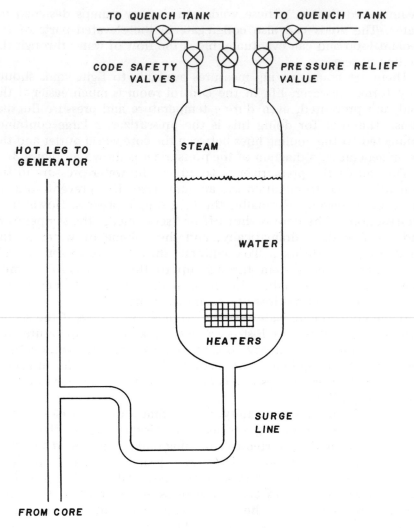

Figure 3. TMI Pressurizer System.

the pressure is to maintain its own steam void, the only one in the system.

The pressurizer is always among the most elevated of the components in the primary coolant loop. Also, there are electric heaters in the pressurizer to insure that it is the hottest point in the cooling system. If steam will form anywhere, it will form here. Thus, a permanent steam void is maintained in the pressurizer, and when the pressurizer is about half filled with water and half filled with steam, the pressure throughout the system is just about right to prevent voids without overpressurizing the equipment. At Three Mile Island, the ideal pressure is set at 2155 psi.

Allowing a steam void in the pressurizer also gives the system extra flexibility in case of a sudden shift in conditions. During overpressurizations, water can expand into the steam void. This gives the operators some time to fight the problem before it becomes necessary to open the pressure relief valve. More important, during a scram, this is the place where the decreasing volume of water causes a growing void. Water can be withdrawn during a shrink all the way to the surge line at the bottom of the pressurizer before a void can escape into the normal loop. In practice, the operators do not like to uncover the heater coils near the bottom of the pressurizer. However, the extra space is available as a margin of safety if needed.

Since the pressurizer is the one place in the primary coolant loop where observable changes are expected during a temperature transient, it is the place where operators tend to focus their attention. It has traditionally been felt that if the pressurizer maintained safe parameters, the rest of the cooling loop would be in good condition. This was one of the primary tenets of the Babcock and Wilcox operator-training program. However, this is one of the principles of pressurized water reactors that is now crumbling.

In addition to the pressurizer, there is one other major component of pressurized water reactors that is foreign to the boiling water design. The primary loop, except inside the pressurizer, is solid water. However, steam is required to drive the turbines. This steam is created inside the steam generator, shown in simplified form in Figure 4.

Hot water from the core enters the top of the generator through a pipe known as the hot leg. Inside the generator, it divides into roughly 10,000 small pipes to maximize the surface area of the pipes. It then returns to the core through the cold leg. Also in the steam generator, but separated by the pipe walls, is the water of the secondary cooling system. It absorbs heat from the primary water through the pipe walls and boils. The steam goes through a set of separators and dryers to assure that only adequately heated steam leaves the generator. This steam turns the turbines, is cooled back to water in the condensor, and returns to the steam generator.

Should the secondary water flow from the turbines be stopped, the available water soon boils away. In the Babcock and Wilcox design, this can happen in less than two minutes. Other designs use different configurations in the generator to gain more than ten additional minutes under worst possible assumptions.

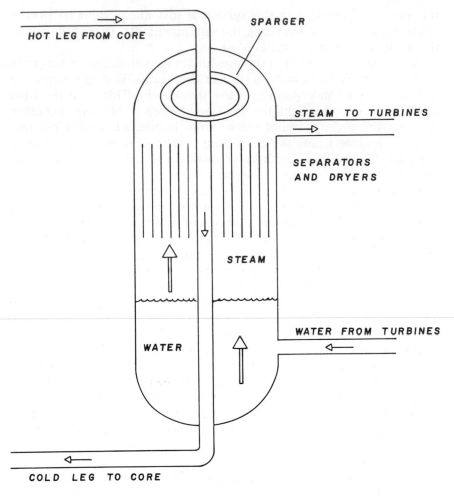

Figure 4. Steam Generator.

However, all this should be a moot point, since there are additional pumps to add water through a different route. At Three Mile Island there are three such pumps, sometimes called emergency feedwater pumps or auxiliary feedwater pumps. After passing through a redundant set of valves, water can be added to the secondary cooling system through a mechanism called a "sparger," which is similar to a sprinkler ring at the top of the generator. In this way, the boil dry time actually tells the operators how long they have to get the auxiliary feedwater system working. Since the auxiliary feedwater pumps automatically engage and reach full power within five seconds of any transient condition, the lead time seemed more than sufficient to Babcock and Wilcox when they created their generator design.

In at least two areas, references have been made to back-up or emergency systems. The safety components built into the reactor now require more attention. As with all reactors, much of the complexity of the Babcock and Wilcox design does not come from the basic components. Rather, the plant is a maze of emergency equipment and redundancy.

This is particularly obvious in the primary coolant loop, where accidents can have severe consequences. There is obviously just one core. However, at TMI, it is encased in an eight inch thick pressure vessel that is rated to withstand the direct impact of a jumbo jet aircraft at 200 miles per hour.[2] The control rods have separate motor drives and will fall into place by gravity if all else fails.

There is also only one pressurizer, since more than one would give confusing results. However, it contains three different relief valves to ensure that at least one will open. Should any open without cause, there are redundant valves to isolate them. The pressurizer also has three independent makeup pumps, any one of which is sufficient during likely emergencies. There are also redundant valves to isolate each of these.

Virtually everything else in the primary system is at least duplicated. The earlier description mentioned a steam generator; there are actually two generators accepting water from separate outlets in the core vessel. Both are normally in operation, although either can be isolated if the need arises. Another major component of the primary cooling system is the reactor coolant pump, a very large and powerful pump that returns water from the cold leg to the core. Ideally, one is needed for each generator. However, there are four of these, with two attached to the cold leg of each generator.

In a loss of coolant accident, there are several emergency pumps available, ranging from high pressure makeup or injection pumps to low pressure residual heat pumps. The high pressure pumps automatically engage when the system pressure falls below 1600 psi, in time to prevent void formation. In addition to the multiple pumps, there are also redundant sources of water, including water with enough boron in solution to stop a chain reaction without the use of control rods.

There is far less redundancy in the secondary system, since the proper operation of this loop is more an economic convenience than a safety factor. The one major safety role of the secondary system is to provide a heat sink in the steam generator to keep the primary system cool. As mentioned, the three auxiliary feedwater pumps pass water through two sets of redundant valves to provide simultaneous flow to both generators or to either separately. At Three Mile Island, these pumps are intentionally separated in the plant and even run off different power sources.

Should the secondary coolant loop be completely clogged, the steam from the auxiliary water could not escape. Actually, the steam escape is regulated to prevent the water from cooling the primary system into a shrink. However, should the regulated release become blocked, it is possible to use an atmospheric dump valve to direct the steam out of the building and into the atmosphere. Given community relations, this noisy alternative is not necessarily desirable. However, it is available and was used during the Three Mile Island crisis.

The rest of the secondary coolant loop relies on bypass valves rather than duplicate components. There is one set of turbines and condensors. The several pumps leading from the condensors to the steam generators have duplicates, but a failure in the one on line will normally cause the flow in the entire loop to stop. This is not, however, the potential problem that it would be in a primary loop blockage, and the safeguards to avoid malfunctions are not as rigorous. If the operators want to use a different coolant flow path in the secondary system because they suspect a problem in a component, the switching capacity is usually available. However, if an on line unit fails, it normally trips the entire secondary system.

So far, it has been most useful to list the components of a pressurized water reactor separately in order to describe the function of each. However, it is now possible to go back to Figure 2 and list the flow path and major safety systems that existed and were in operation at the Three Mile Island nuclear plant on the morning of March 28, 1979. This is a description of the reactor as it was supposed to operate and was operating before the difficulties began. The next chapter recounts the reactor trip that turned sour.

To start such a summary, one additional piece of information is needed. Like many American reactors, Three Mile Island contains not one, but two reactors. On the island there are two cores, two turbine systems, and two sets of cooling towers. This is not particularly unusual, but most multiple reactors share a common control room. Three Mile Island does not. Each unit at TMI has a separate control room, and the reactor systems operated by each control room have several differences. This is why control room operators are licensed by the NRC for one or the other unit, but not for both without separate tests. One advantage of separate control rooms is that each has an emergency headquarters for the other unit in the rear of its control room, should this become useful. This is unusual and highly redundant, but served its purpose in late March 1979.

Leading into the crisis of early March 28, the system was operating roughly as intended. Unit I was shut down for refueling, but Unit II was operating normally. The Unit II core was generating 97 percent of its rated heat capacity, heating the water to an average of a little over 580°F and 2155 psi pressure. The heated water rose to the top of the two steam generators through the hot leg pipes. Because of their shape, these pipes are also commonly called the candy canes. The pressurizer was attached to the candy cane leading to steam generator A. On reaching the tops of the generators, each set of primary coolant pipes split into approximately 10,000 small tubes to be cooled in the steam generators. On leaving the generators through the cold legs, the water passed by the let-down system designed to let water out of the loop if necessary. One valve was partially open for normal maintenance, as described in the next chapter. From the cold legs, the water entered one of the large reactor coolant pumps to be forced back into the base of the core.

The secondary coolant steam from the generators was separated by the dryers and separators so that only the superheated steam escaped. From there it flowed through the high pressure turbine and the two low pressure turbines. It then cooled back to water in the condensor, where the steam transferred its heat to a third cooling loop attached to the cooling towers. At the end of the condensor was a polishing system to remove unwanted minerals and contaminants from the water.

At the tail of the condensor was a series of pumps to draw water through the condensor and then rush it back to the steam generator. In simplified form, these included the condensate booster pumps and then the condensate pumps and then the main feedwater pumps. Within the steam generator, the cycle began again.

In one of the ironies of history, at 4 A.M. on March 28, 1979, Unit II was within twelve minutes of its first anniversary of going critical. While some of the crew on duty had been on hand for the earlier event, the people in the control room had forgotten the coincidence. In that year, the unit had suffered more than its share of growing pains and had been shut down for sizable periods. Some of those problems were listed in Chapter 1. However, no one had publicly guessed that this plant might be more susceptible to problems than most of the others in the industry.

NOTES

1. Testimony of Herman Dieckamp before President's Commission on the Accident at Three Mile Island, May 30, 1979, p. 17. Testimony of Edward Frederick before President's Commission, May 30, 1979, p. 205.

2. Testimony of Richard Dubiel before President's Commission, May 31, 1979, p. 196.

Chapter 3

A Transient Goes Awry

As is typical for nuclear power stations, the two control rooms at TMI were busy but not particularly hectic places in the early morning hours of March 28, 1979. Every few minutes, annunciator beepers would draw attention to flashing lights on the walls surrounding the control rooms. One of the two operators on duty would immediately notice the cause of the alarm. However, the annunciators normally signaled minor malfunctions or simple expected shifts in the unit's status and were handled or acknowledged by the operators with no particular sense of urgency or delay. With all the optional flow paths and manual override capabilities in the unit, there was no clear reason to treat the routine annunciators as threatening or anything other than ordinary.

One indication of the flexibility of the entire unit was the scattering of tags tied to levers around the room marking various systems as inoperative. During the day, regular maintenance shifts would trace down and repair some of these malfunctions, concentrating first on the more important ones. However, some systems were too peripheral to the function of the unit to deserve any sense of priority. Many malfunctions were in-line or in "hot" areas and could not be repaired until the next scheduled reactor shutdown. As a result, some of the tags had been there for months. These continuing malfunctions remained on the annunciator boards as solid rather than flashing lights, without activating the beeper system. One shift supervisor commented that the fewest lights he had ever seen on at one time was fifty-two.[1]

The maintenance crews during the day were a help to the operators in more than one way. They repaired and checked various safety systems. They also kept the place busy and even interesting for the day shift with their constant level of activity. However, the main distraction for the night crew was "switching and tagging," during which an operator would take various readings, testing equipment and tagging any items needing attention. This was a welcome assignment, relieving the operator of the normal routine of checking the annunciators that constantly interrupted the seemingly endless administrative log keeping and paperwork.[2]

In the early morning hours of March 28, the adjoining Unit I of TMI was in a state of hot shutdown. The unit had just completed the awkward and time-consuming process of refueling and was using heat from Unit II's steam generators to slowly bring the system up to temperatures where it could be started without excessive shock to the cooling system.[2] Unit II was operating at its normal 97 percent of capacity, with control rods partially inserted to equalize the heat across the core.

The control room operators were completing their fifth consecutive night on the 11 P.M. to 7 A.M. shift. As usual, Unit II's crew of four consisted of two operators, a shift foreman for Unit II, and a shift supervisor for both Units I and II. While this shift clearly had the least desirable hours of the day, a rotation policy assured that it did not become burdened with the newest or least influential personnel. Rather, while the company called in its top operators and executives later in the day, some of the personnel on duty at 4 A.M. remained instrumental in running the control room well into the evening. Their training, experience, and trust by the company was considerable by industry standards.

Of course, all were licensed by the NRC and had been through six months to a year of training, followed by lengthy oral and written examinations on the operation of Unit II at TMI. As shift supervisor for Units I and II and a senior operator, William Zewe had begun his training with six years as a reactor operator for the Navy.[4] He had been working at Three Mile Island for seven years, beginning as an unlicensed auxiliary operator and working up through shift foreman and then shift supervisor. Since he held the NRC senior operator's license, Babcock and Wilcox had permitted him to go through their training and refresher course at least annually for the past few years. The course included twenty hours of classroom lectures and questioning followed by twenty hours on the control room simulator in Lynchburg, Virginia. Simulators are not required for NRC licensing, and many companies do not have one. However, all TMI operators practiced on the simulator at least once every two years. Senior operators were scheduled for the simulator at least annually. Zewe had most recently been through this experience with the Unit I operators in January 1979. The other Unit II operators had completed this course most recently in July 1977. Ironically, they were scheduled to return in mid-April, two weeks after the accident. James Floyd, Supervisor of Operations for Unit II, was attending a course in Lynchburg on March 28.

The simulator training concentrated on emergency procedures that could not be easily practiced on an operating reactor. Babcock and Wilcox had programmed into the simulator what their experiences told them was the logical consequence of each possible action in the control room. The simulator could be set for certain emergencies or malfunctions, and the operators could fight the problems with continual indicator feedback on how they were doing.

However, none of the operators had practiced the events that were to transpire in Pennsylvania on March 28. Had the simulator been presented with the problems they were to face, it would not have been programmed to respond.[5] Experienced and repeatedly trained operators were about to encounter readings from their instruments that they would have thought impossible.

At 4:00 A.M., Senior Operator Fred Sheimann was acting as shift foreman for Unit II. Being in charge of operations throughout Unit II, and not just in the control room, he was in the turbine building in the general vicinity of the condensate pumps.[6] Operator Craig Faust was just completing a switching and tagging exercise on the control room indicators for the main turbine generator.[7] Shift Supervisor William Zewe was in his office at the rear of the control room for Unit II.[8] The second operator in the control room was Edward Frederick. Seven additional nonlicensed auxiliary operators were at various stations throughout the plant.[9]

Unit II was almost running itself through its automatic mode, called the Integrated Control System (ICS). The reactor coolant system was normal. However, the operators were overriding the automatic system in two areas to accomplish specific tasks.

First, the spray control valve that allows makeup pump B to add water to the pressurizer was being held open manually. Boron, which is added to the core coolant water to regulate the rate of fission, was disproportionately concentrated in the pressurizer. Water was being added to the pressurizer to try to flush the boron out into the rest of the loop. Boron is a very effective regulator of the chain reaction, but not when it is unevenly distributed in the system.

Apparently, the boron problem was caused by one of the three valves on top of the pressurizer. Either the pressure relief valve or one of the code safety valves was leaking up to six gallons per minute, but there was no way to fix it or even to get near enough to identify the problem valve before the next shutdown.[10] Since six gallons per minute was easily made up with makeup pumps, and since any of the valves could be isolated in a malfunction, there seemed to be no reason to shut down early to repair the valve.

The other activity was in the turbine building, near where Shift Foreman Fred Shiemann was checking. The secondary loop water flowing through the turbine naturally picks up unwanted minerals. As the steam is condensed back to water, it is run through a polishing system where minerals are absorbed by pellets coated with a resin. When the resin becomes saturated with minerals, which happens after about twenty-eight days of use, the pellets from any of the eight chambers can be shifted to a regeneration receiving tank where they can be "fluffed" or cleaned and returned to service. The resin is moved by demineralized water pressure and fluffed by air pressure.

At 4:00 A.M. on March 28, resin was being transferred and fluffed. The sequence of events inside that polishing system for the next half minute was not known to the operators or anyone else for a few weeks, but within moments it led the entire unit into trouble. While being transferred to the regeneration tank, the resin pellets blocked the line. Water pressure used to push the resin exceeded the air pressure fluffing the resin. A check valve on the air system ensures that water does not flood the air system during such a pressure imbalance. Because of previous malfunctions, Met Ed has also installed six traps in the air line to stop moisture. However, now that the back-up system was needed, the check valve failed to seat properly and leaked in excess of five gallons per minute, flooding the air system.[11] This and continuing malfunctions are shown in Figure 5, p. 44.

To protect the rest of the secondary water loop when there are problems in the polishing system, there are valves at the entrance and exit to isolate the polisher. Unfortunately, these valves are activated by the same air pressure system that had just become flooded. Slowly and steadily, and without the knowledge of anyone in the control room, the water in the air line was now pushing the valve controls to isolate the polishing system.

In Unit I, a polisher bypass valve would automatically open when the polisher became isolated. In Unit II, the controls to isolate the polisher and to open the bypass system were separate.[12] As a result, the closing isolation valves were about to block the entire secondary cooling system without the knowledge of the control room operators.

These circumstances can be made to sound more dramatic than they warrant. Line blockage and flooding of the air lines had both occurred before, and William Zewe had complained in writing on May 15, 1978, that "resultant damage could be very significant."[13] The most serious consequence of blocking the secondary cooling system is that the reactor may have to be tripped off or "scrammed." In the TMI system, even this can theoretically be avoided, although it usually is not.[14] A reactor trip is a fairly traumatic event for the equipment, involving sudden severe temperature and pressure fluctuations that require numerous counteracting measures. Besides, it throws 1000 megawatts of heat out of energy production, which is not welcome news to the utility company. However, both the equipment and the operators are prepared to handle scrams, and they are not that infrequent in the industry.

With the polisher isolation valves closing, a bothersome but for the moment unthreatening chain of events began. First, the condensate booster pumps tripped off at 4:00:36 A.M. These create a suction at the tail of the condensor to help convert the steam back to water. However, they are sensitive to their role as a suction generator, and they have tripped at TMI several times before when encountering difficulty.[15]

Within a second, both the condensate pumps and the main feedwater pumps had tripped downstream from the booster pumps. All this happened far too fast for the operators to intervene, so the Integrated Control System proceeded as programmed. Accordingly, the turbine was automatically tripped at 4:00:37 A.M., having lost its steam supply from the main feedwater pumps. The entire secondary system was now grinding to a halt.

Craig Faust had just finished a switching and tagging assignment and was turning to face the main control panels and talking to Edward Frederick when the annunciator sounded.[16] In turning around, he caught sight of several flashing lights on the Integrated Control System panel and realized that they were not routine. He pointed at the panel and told Frederick that something was wrong with the plant. Additional lights were almost immediately added for the turbine trip and an automatic reactor "run-back."

The economic function of the malfunctioning secondary cooling system is to drive the turbines for electricity. However, equally importantly, the secondary system absorbs the tremendous excess heat that the core generates in the primary coolant loop. With the secondary water no longer flowing, several automatic steps were necessary to make up the lost heat sink for the primary cooling system.

One adjustment was to drive the control rods part way into the core to "run-back" or reduce reactor power. This way, less heat was generated in the primary system, and the reactor could hopefully be saved from a complete scram. However, this process was slow, and more immediate steps were needed to absorb heat.

By industry standards, the Babcock and Wilcox steam generator design holds a relatively small amount of secondary cooling water in the steam generator. With loss of the feedwater pumps, the secondary cooling water could boil away in less than two minutes.[17] Therefore, auxiliary feedwater pumps immediately engaged to add more water to the secondary coolant loop. That way, as water in the secondary system accepted heat from the primary coolant loop and boiled away, the auxiliary feedwater pumps would be able to add more water directly into the generator. The main feedwater pumps added this water during normal operation, but they had just tripped.

In accordance with their emergency procedures, the operators watched the instrumentation to verify that the control rods had been driven and that the three auxiliary feedwater pumps had started. Also, William Zewe heard the commotion and looked up from his paperwork and through the glass office wall to see the ICS lights flashing.[18] He jumped out of his chair and hurried into the room in time to see that the turbine had tripped and that the operators were at their assigned posts to monitor the transient in the secondary cooling system.

It had been about two seconds since the annunciators had sounded in the control room. However, without the knowledge of the team monitoring the equipment, the transient and the plant were already in serious trouble. The three auxiliary feedwater pumps that automatically engaged to replace the main feedwater pumps were supposed to push water to the generator through two pipes, each controlled by a valve, designated EF-V (emergency feedwater-valve) 12A and 12B. These valves were operated from the control panel by switches and red-green light indicators. They were to be left open at all times except when being tested. In fact, there was even a bypass for each valve, with additional valves that were to be open if the first set failed. As this event developed, all of the valves were closed. The auxiliary pumps engaged, but the water they tried to pump was blocked from entering the generators, where it was now needed.

Ironically, these valves had been tested using the procedures in the Plant Operation Review Committee manual on March 26, less than two days before the accident. At that time, Shift Foreman Carl Gutherie had requested control room operator Earl Hemmila to assemble a group to check the entire auxiliary feedwater system.[19]

Hemmila closed the valves from the control room. The team then tested the valves, pumps, and coolant supplies. At the completion of the test, operator Martin Cooper opened the valves and verified along with Hemmila and one other operator that the lights had switched to open.

Somewhere in the intervening forty-two hours, the valves were reclosed. That reclosing has become the subject of intense speculation, including such charges as incompetence and possible sabotage.[20] Even though the valves have been inspected repeatedly since the accident, there is no physical evidence available at this time to explain the closing. At this point, there are only possibilities. For instance, valves have been known to vibrate shut. However, the chances of this happening on two separate but related valves are extremely remote. There were no obvious malfunctions in the valves; their control room indicators were correct, and they opened on command. They could have been left closed by the March 26 inspection team, although three people claim to have witnessed the reopening. The valve switches are near each other on the control panel and could have been changed by operator intent or error. However, the motive would not be clear since this action alone would aggravate but not endanger a transient. Finally, there is a set of remote switches in the plant, and they could have been intentionally tripped. The switches on the valves themselves were padlocked.

In the President's Commission summary and technical staff analyses, this mystery remains unsolved and will continue to generate controversy. It will not be cleared up here and may never be. However, the fact that the control room lights showed the valves to be closed and that no one noticed indicates something unsettling about the nature of control rooms. However the valves came to be closed, they would not have been a problem if the sophisticated computer system in the control room had alerted someone.

The control room has literally thousands of indicators and lights. Considerably more information is in the computer available for display, but is not readily visible. At any time, some of the visible lights on a control panel should be red and some should be green. In other words, a light indicating a valve out of position is not obvious in a visual scan. Information in the computer may not even be visible.

A new operator coming on duty could take the time to look at each light and reading in the computer and analyze its information. However, there are several duties to perform, and these would have to be delayed during the leisurely survey. Therefore, the operators have developed time-saving techniques to analyze what information is important. When a shift superintendent takes over the log, he reviews the major actions recorded since his last shift. Remembering the status of the reactor when he was last on duty, he makes a mental note of the changes. Also, operators routinely pass informal turnover sheets to the next shift, listing any useful information not in the log.

Operators on the floor also develop time-saving techniques to keep up with the system's status. Every few minutes they scan the control panels, looking for anything that does not appear normal. With annunciator panels to warn of important system changes, these procedures are usually sufficient.

In fact, this scanning technique has occasionally caught problems such as valves out of position before they affected the operation of the plant. Finding valves out of alignment is not a common occurrence, but it does happen. Station Manager Gary Miller had participated in five of the seven valve alignment investigations occurring since 1973 at Three Mile Island.[21] One of the valves so discovered was on the high pressure injection system and could have caused potentially catastrophic results compared to those on the auxiliary feedwater system.

Whatever the cause or the limitations of the indicator layout, the fact remains that the valves isolating the auxiliary feedwater system were shut at the beginning of the accident and remained undiscovered for eight more minutes. Not knowing this, Faust verified that the pumps had started. There was no gauge to indicate that the pumps were actually moving water. Instead, to verify flow, Faust looked at the steam generator water level indicator, which was temporarily sufficient since the boiling process had just begun.

Of course, any of the three people in the room could have looked at the indicators for the closed valves. However, to check the switches they would have had to ignore the annunciators and established emergency procedures during what they knew to be a serious transient. It was now two to three seconds after the annunciators had sounded, and perhaps fifty lights were flashing. The operators were working on trained reflex and did not yet understand the exact nature of the malfunction.

When the secondary cooling system can no longer absorb heat from the primary loop, it is normal for the pressure and temperatures in the primary system to climb dramatically. The pressure relief valve above the pressurizer drains the excess primary water to avoid damage to the system. If the overpressurization threatens to become dangerous, the code safety valves open at a higher pressure (about 2500 psi) to insure that the primary system cannot be ruptured or damaged. These valves are more critical on Babcock and Wilcox plants than on most other pressurized water reactors since the water reservoir in the steam generators for secondary cooling is so small. In fact, some manufacturers do not even include pressure relief valves and rely solely on the code safety valves for a real emergency.[22]

However, Babcock and Wilcox designed the pressure relief valve to function as part of the normal turbine trip procedure. It is set to open automatically at 2255 psi, which is a rise of just 100 psi over normal. With any luck, the combination of the run-back reactor, the open relief valve, and the auxiliary feedwater pumps can prevent the serious overpressurization that forces a reactor trip. Usually, however, these systems cannot stop the pressure rise in time.

If the pressure reaches 2355 psi in the primary system, or 200 psi over normal, the reactor automatically trips to remove the source of the heat. If this happens, the primary cooling system then "shrinks," and new water must be added quickly to avoid having all of the water drain out of the pressurizer.

Three seconds into the event, annunciators indicated that the pressurizer level and pressure were climbing rapidly as expected. At six seconds when the system climbed past 2255 psi, the pressure relief valve opened as designed. With luck, the heat sink and open valve might stop the pressure rise and stabilize the system. However, the emergency feedwater pumps were not creating a sufficient heat sink in the generators, since they were blocked by the closed valves and not pumping water to the generators at all. Therefore, at eight seconds into the event, the primary coolant loop passed 2355 psi, and the reactor automatically tripped. As Faust watched the panel, he was the first to notice that the "control rod bottoming" lights were coming on. Zewe, who had arrived a few seconds into the transient and was now scanning the boards, spotted the reactor trip next and announced it to the room.

With the reactor tripped, all the thinking reversed. As the chain reaction stopped, the core would drop 93 percent of its heat generation in about one minute. Instead of the earlier rapid pressure and water level rise, the water would now reverse itself and shrink rapidly. In fact, there was the potential danger of pulling all of the water out of the pressurizer and forming pockets of steam in the coolant line. Of course, this is a potential danger in any pressurized water reactor trip, and the makeup pumps are designed to add thousands of gallons of water per minute into the cooling loop if needed to offset shrinkage. The shrinkage process even takes a few seconds to start, permitting the operators to verify that the necessary actions have been taken. Therefore, reactor trips or scrams are generally considered much more of a nuisance and embarrassment than a safety hazard.

With this few second break in new developments while waiting for the shrink, the operators were able to initiate a few deliberate actions to help the system prepare. Craig Faust shut the one let-down valve that had been slowly releasing water from the cold leg of the loop.[23] As mentioned earlier, makeup pump B had been in continuous operation to equalize the boron concentration and to provide water to replace various leaks and to seal the pump shafts. This let-down valve had been the escape route for the extra water, but all of the water in the primary loop was now needed to offset the expected shrink. Also, Faust attempted to start makeup pump A as a supplement to the already operating B. There is a time delay on the switch, and he let go too soon. Edward Frederick tried it, and the pump started.

Zewe began following his emergency procedures for a reactor trip.[24] He went to the public address system and announced to the auxiliary operators in the plant that there had been a reactor trip. Fred Sheimann, who heard the loudspeaker announcement and could hear for himself that the secondary system had stopped, now headed for the control room. The other operators headed for their assigned posts or to safe areas.

Edward Frederick watched the boards as required to verify that all the control rods had seated properly.[25] While doing this, he checked the neutron count rate as a second indication that the chain reaction was ending. He also flipped the pressurizer controls from manual to automatic to allow the computer to automatically manipulate the controls as needed.

It was now thirteen to fourteen seconds after the first annunciators, and while the reactor trip was definitely the excitement of the morning, the sequence seemed normal. According to their drilled procedure, Edward Frederick now stationed himself at the part of the control panel where he could best monitor the primary system, and Craig Faust walked over to monitor the secondary system. They could then pass any pertinent information verbally. Among their first actions was to try to inventory exactly what was still operating.

Faust on the secondary system noticed that the generator water level was falling as expected since the auxiliary feedwater pumps took a few moments to have an effect. He also verified that the throttle valves on the auxiliary feedwater pumps were now opening to increase feedwater flow to the generators and to restore the proper water levels. Both of these readings were normal, and there was no reason for Faust to suspect that the flow to the throttle valves was actually blocked.

As a distraction, Faust then noticed that while the entire secondary loop was now off-line, one of the eight valves that should have automatically closed to isolate the turbines was hanging up. He pushed the turbine trip button to reactivate the valve, but it still stuck. In these first few seconds, the operators did not want any generator steam escaping to the turbines or anywhere else, since the trapped steam would keep the generators from overcooling the primary system and causing a larger shrink. Instead, the turbine isolation valves were supposed to maintain pressure at 1010 psi. Ironically, Met Ed was later to discover that this valve was in fact closed, but it had slammed shut so quickly that it had severed the indicator arm.

Frederick on the primary system was directing his attention to the two focal points of his training. He continued to verify that the chain reaction was stopping and that the pressurizer level and pressure were within acceptable limits.

Then, with no fanfare, the second and far more serious failure occurred. What had been developing into a serious transient with the blocked auxiliary feedwater pumps now set itself on the course of becoming a potentially disastrous one. While the pressurizer water level was still within two seconds of peaking, the more rapidly changing pressure was dropping dramatically. It passed the 2205 psi setpoint for the pressure relief valve to close. With expected shrinkage, the open valve would now be a serious hindrance to maintaining adequate water levels in the pressurizer. The light in the control room indicated that the valve closed. Actually, it remained jammed partly open.

The operators understood that the light showed only that the electrical circuits had withdrawn the command to the solenoid to hold the valve open. Normally, that was good enough. However, if the circuit or the solenoid malfunctioned, the lights would incorrectly say that the still open valve had closed. The pilot-operated relief valve design has a poor history, having failed once before at this plant and eleven times throughout the industry.[26] Therefore, during a transient, operators often double check the accuracy of the lights.

If the operators are suspicious about the valve signal, there are two back-up indicators available. First, each of the three valves atop the pressurizer is attached to a pipe that directs its discharge to the quench tank. A thermocouple, or electronic thermometer, is strapped to the outside of each of these pipes to read the temperature of the discharge. While there was not a gauge in the TMI control room for this thermocouple, its reading can be called up on the computer at any time. As a second backup, there is also a thermocouple on the quench tank further downstream. Later in the transient, Zewe asked for both readings.

The readings on the first thermocouple require additional explanation because there was some confusion in the control room during the first couple of hours on what the readings indicated.[27] Ideally, when the pressure relief valve is closed, temperatures at the thermocouple should be at the containment building's room temperature of about 100°F. But a system of 2155 psi and 582°F is destined to leak, and thermocouple readings of 130°F are rated as normal. In fact, leakage in this valve had been a problem since the unit opened, and temperatures of 175–195°F were rather standard.[28] At 200°F the thermocouple activated an annunciator, and Zewe had been on duty before when it had done this with normal leakage.

Over the next two and one-half hours, the first thermocouple had readings that began in the 290s and dropped to 228°F. Obviously, these readings were not in the normal operating range. However, neither were they in the 300–600°F range that the thermocouple can read when the valve is fully open. As the operators encountered these confusing midrange temperatures over the next two and one-half hours, they decided that the pipe must be holding some residual heat from the earlier release and that the temperature would soon fall to normal leakage levels.

It was now fifteen seconds after the turbine trip and seven seconds after the reactor trip. The previous water rise was now over, and the primary system had entered the phase when pressures, temperatures, and water levels would all fall rapidly. This is a delicate period when it is potentially dangerous to inject too much or too little water, so the operators watch the system very carefully. Of course, this cycle is normal in scrams, and the makeup system can handle it with regularity. Given the desire to avoid uncovering the heater coils in the pressurizer, the operators can even help the process along by starting some pumps early. Accordingly, Frederick engaged another makeup pump to keep the pressurizer level above the desired minimum of 100 inches, with an intent to stabilize the level eventually at 200 inches. The maximum possible level in the pressurizer is 400 inches.

Beginning at seven seconds and ending at forty-six seconds after the reactor trip, the pressurizer water level fell. From a high of 255 inches, it plummeted to 158 inches. Then it bottomed out and began to rise again. Obviously, the water had not been shrunk out of the pressurizer; it had not even come close. There was a brief feeling of satisfaction among the operators. They had been repeatedly trained to counteract the effects of a poor scram, but by the criterion of pressurizer water level, this one had been relatively easy.

But "satisfaction" was far from the feeling that the event was over. During the water level shrink, new annunciators began to light up, giving messages that simply did not make sense. While they still had no idea what problems they were facing, it was becoming increasingly clear to the people in the control room that something was seriously wrong.

One of the problems was a mere aggravation. Makeup pump 1A, one of the two pumping water into the primary loop, tripped off during the shrink. Frederick restarted it, but the pump's control room switch was breaking down, and it tripped a few more times during the morning.

More ominously, the water in both generators lit annunciators when the levels fell below the low water setpoint of thirty inches. The water was boiling away faster than the auxiliary feedwater pumps could replace it. Of course, Faust did not realize that the water from the pumps was completely blocked. Instead, he drove the throttle valves further downstream to the full open position to speed the water flow to the steam generators. However, there was no water to flow.

Almost a minute into the event, the attention in the room was drawn away from Frederick's apparently stabilized pressurizer to the growing generator problem. Had the operators now called up the reading from the thermocouple outside the pressure relief valve, they would have received a reading of 294° F, the high of the morning.[29] However, of the perhaps one hundred annunciators that were now lit, the ones for the generators most clearly indicated that there was a crisis in the making.

The water in the two generators had reached the lowest reading that could be recorded on the level-indicator probes—a level of ten inches. Then the water fell below this level, and the generators appeared to be dry. With all the water being pumped into the generators from the auxiliary feedwater pumps, this made no sense. So Faust began looking for other indicators to check his best guess that perhaps the water level sensors were wrong.

As the primary coolant pipes ran through the generators to be cooled, their temperatures were measured at the entrance (hot leg) and at the exit (cold leg). The differential measured how effectively the generators were cooling the primary water. For all practical purposes, the differential reading showed that the generators were not cooling the primary system at all. Another available reading gave the average temperature of the primary water supply. It was rising, once again indicating that the heat sink had been lost. Finally, the pressure in the secondary cooling system was rising, indicating that new water was not keeping the temperatures—and therefore pressures —reduced. With all these indications, there could be no question that the generators were in fact dry.

But why? With hindsight, it is easy to argue that Faust or Zewe should have checked the controls for the auxiliary feedwater system from the water supply to the generator until one of them found the problem. Even with foresight, this is not an unreasonable argument, although over ten feet of control panel with several rows of controls were involved. But during this period shortly after one minute into the event, they did not check these controls. In testimony before the President's Commission, the operators offered no explanation for not checking the valves at this point other than that they did not think of it. And there are at least three reasons why they may have encountered difficulty reasoning through this confusing situation.

First, Faust had just forced the throttle valves open. These are large valves that take a few moments to respond. Also, the level sensors in the generators were spaced in such a way that a change in water level would not be noticed until it became substantial. Patience was not easy at this point, but it was required to tell the effects of the previous action.

Second, there were in excess of one hundred annunciators flashing, with new ones going off in excess of one per second. All the operators now maintain that the emotional atmosphere in the control room remained as calm as the circumstances would permit. However, as Faust commented later, "I would have liked to have thrown away the alarm panel. . . . It wasn't giving us any useful information."[30]

The third and most disturbing problem was that the pressurizer began to act irrationally. Once the shrink was stabilized, the pressure and the water level should have risen in the pressurizer to a safe working margin. This reflected the sudden stabilization of core temperatures and the fact that the makeup system was still working. It was an expected and useful rise that Frederick intended to try to stabilize at about 200 inches. However, the pressurizer water level almost immediately began to rise at what seemed to be impossibly rapid rates. From about one minute to almost five minutes into the event, the pressurizer overcame all attempts to slow the water level rise.

Under normal conditions, a water rise in the pressurizer would be contained by the steam trapped above. But the pressure relief valve atop the pressurizer was open, and water was able to rush into the pressurizer faster than the operators would have thought possible. Since the pressurizer was designed to control the pressure level in the rest of the primary coolant loop, the first priority was to stabilize it.

While attempts were begun to stabilize the pressurizer, a critical stage was passed in the primary coolant loop. Under normal circumstances, water did not boil in the primary loop, despite the high temperatures, because the pressure was high enough to prevent boiling. However, if the temperature rose high enough or the pressure dropped low enough, the water would boil in the pipes and form steam voids. Just as steam was designed to hold down the water level in the pressurizer, it would resist water flow through its voids in the cooling lines.

The primary coolant temperature was slightly elevated because of the loss of the secondary coolant heat sink. More importantly, the release through the pressure relief valve was causing the pressure to drop rapidly. It was six minutes into the accident before the saturation temperature-pressure combination was reached and steam began to form throughout the primary cooling system. However, the pressure and temperature were not distributed evenly throughout the loop, and about ninety seconds into the event, voids began to form at some points and to slowly collect in the top of the reactor vessel and in the hot legs. The operators were able to read the pressure drop in the system, but the water level in the pressurizer was rising very rapidly. These two conflicting readings made no sense to the operators. While the operators were not panicky, they were completely perplexed.

Somewhere in this period, it occurred to Zewe that the pressure relief valve might be stuck open. Rather than ask for the thermocouple reading just outside the valve, he asked about the thermocouple that was located in the reactor coolant drain tank, or quench tank. The gauge for this reading was behind a distant control panel, perhaps a forty foot walk from the pressurizer readings.[31] Some operators consider this reading more reliable than the one just outside the valve since it gives some notion of the total volume of steam released. Unfortunately, because the valve had been stuck open only a short time, the reading was still in a reasonable range, somewhere in the neighborhood of 100° F.[32]

The contradictory feedback from the indicators continued to get more confusing. Water in the pressurizer continued to rise, although for the first two minutes the rise was needed, and no moves were made to fight it. But shortly after two minutes into the event, the pressure in the primary cooling system fell below the setpoint of 1600 psi, and the Emergency Core Cooling System (ECCS) came into automatic operation.

When the ECCS engages, it is an unmistakable warning that there is serious trouble. Reactors did not begin to include this system until

1966, and the last one was added to an existing plant in 1972. It is the final and extremely powerful automatic back-up system in a reactor. If all else fails, the controls of the ECCS can pour thousands of gallons of water per minute into the core vessel with immediate access to perhaps one million gallons of reserve water.

The ECCS program in the computer overrode Frederick's manual control of the two makeup pumps and forced them wide open, adding 1000 gallons of water per minute into the primary loop. It started the decay heat removal pumps and the third makeup pump and opened the throttle valves on all of these wide open.

About two and one-half minutes into the event, Shift Forman Fred Sheimann entered the control room to see what had happened. Having no previous hint of irregularities, he encountered the Emergency Core Cooling System pouring well in excess of 1000 gallons per minute into the primary system, low primary loop pressures, dry generators, and a dangerously rapidly climbing pressurizer water level. In the midst of all of this nonsensical information, there was no time nor apparent benefit in reasoning through the problem calmly. What was happening simply could not happen. This was the time to fall back on emergency training.

Sheimann pulled out copies of the "Emergency Procedures for Turbine Trip/Reactor Trip" to use for reference. However, he was quickly diverted to the trouble spot of the room, the pressurizer level. There was no serious thought now that there might be an open valve in the primary system. If the rapidly rising water level indicated anything, it was that the loop was watertight and that the high pressure injection pumps were filling it too quickly. Of course, this interpretation assumed that the pressurizer was operating normally, which it was not with what amounted to a hole in the top of it.

There was a short discussion and agreement that the water rise in the now two-thirds full pressurizer had to be stopped, even if it required the highly irregular move of overriding the Emergency Core Cooling System.[33] Then, shortly after three minutes into the transient, the operators as a team began to act. Frederick and Sheimann handled the controls on the primary system while Faust stayed with the dry generators and Zewe tried to coordinate activities.[34] Frederick pushed the six bypass buttons necessary to override the Emergency Core Cooling System. They then turned off one makeup pump and throttled the others to cut the flow to the core to one-half of its previous volume.

They expected a major drop in the rate of water rise in the pressurizer. It faltered for a moment, but then resumed its old rate. At four and one-half minutes into the transient they tripped another makeup pump and throttled back the other as far as it could go without damaging the pump seals. The pressurizer was now almost full and steadily rising.

Shortly after five minutes into the event, they reopened the let-down valve that had originally been open in the cold leg and began to drain the primary loop in excess of 140 gallons per minute. Since it was in fact a second hole in the primary system, the let-down valve caused some erratic behavior. The pressurizer level fell dramatically but only momentarily. Also, because of the voids in the system, the flow in generator A decreased, and the coolant temperature began to rise. Like virtually everything happening now, these events did not seem logical to the operators. Combined with that, the level indicator for the pressurizer began to slowly bounce because of the unprecedented high readings, giving confusing information.[35]

With the fall in the pressurizer level, the operators felt compelled to throttle the let-down system so that it let out only about seventy gallons per minute. The pressurizer water resumed its rapid ascent. At six minutes into the event, the water level reached the maximum 400 inch level. There was no steam left in the pressurizer. Frederick announced to Zewe, and therefore to the room, that the plant was now "solid."

A solid plant is a true anomaly in pressurized water reactors. The operators are instructed to never drive all of the steam out of their systems. There is no training to handle such a condition since it is theoretically so easy to avoid a solid plant. As one indication of industry faith in the procedure, Frederick, Faust, and Sheimann had never seen a solid plant in reality or in the simulator.

Zewe had intentionally helped run Navy reactors solid three times to test specific equipment.[36] However, Zewe had never seen anything like this. The core temperatures were running relatively hot, though not dangerously so. At this temperature, a solid plant would require reasonably high pressures to avoid natural steam formation. But the core pressure was at an extremely low 1200 psi and falling. Under these conditions, steam should have been forming rapidly, but the pressurizer was solid.

Of course, water was flashing to steam throughout the system and congregating in the elevated and hot portions of the coolant loop. However, any steam that formed in the pressurizer immediately bubbled up and out of the pressure relief valve. As a result, the plant looked solid when in fact it was largely blocked by steam pockets.

Frederick and Sheimann continued to fight the full pressurizer for the next two minutes, trying varying speeds on the makeup and let-down systems. Some actions seemed to work for short periods, but actually the level indicator in the full pressurizer was continuing to bounce, giving false hope to the operators. The quench tank, which was now receiving water as well as steam from the open pressure relief valve, reached capacity. It is a sizable tank, but it is designed to withstand only about 150 psi before venting its contents. It reached 150 psi when the pressurizer went solid and began leaking onto the floor of the containment building.

Leaks in containment are normal, caused primarily by condensation and the impracticality of installing watertight seals on components given their operating requirements. When enough water collects on the floor of containment, a sump pump automatically engages to drain it away.[37] Since there is often some low level radiation in the water, it is sent to airtight tanks in the auxiliary building for later decontamination and disposal.

There are three such tanks available in the auxiliary building at Three Mile Island. The auxiliary building sump tank was out of commission, since it had blown a rupture disk designed for emergency pressure release, and was no longer airtight. There is also a tank commonly known as the auxiliary building sump. The sump pump was aligned to a third tank, called the miscellaneous waste hold-up tank.

At seven and two-thirds minutes into the transient, the sump pump started. However, while the valve alignment controls clearly showed that the sump pump was connected to the miscellaneous waste hold-up tank, the level indicator on that tank did not move during the accident. It is now known that the sump tank, the one with the blown rupture disk, began to fill instead.[38] As of this writing, no one has been able to reenter that part of the auxiliary building to determine why this happened. However, the still very mildly radioactive core cooling water now had an open route through the pressurizer into containment and out the sump pump to the floor of the auxiliary building. Should this water become radioactively contaminated, it could release its radiation to the atmosphere through the air-conditioning vent system. Fortunately, that had not yet happened.

Back on the secondary system, it became obvious with time to Faust and Zewe that the open throttle valves were not solving the problem of the dry generators. Faust began looking for other causes. For a short period of time, he took manual control of the automatic pressure release for the secondary generator steam since the pressure readings were confusing. However, while he managed to control the releases, it did not help his dry generator problem.

Approaching eight minutes into the event, Faust decided to retrace the entire feedwater system across the control panel indicators.[39] He leaned across the control panel to verify that the pumps were running, ironically blocking his vision of valve EF–V12A with his body. He scanned across the panel consciously reading each indicator. As he reached the part of the panel near him, he noted again the yellow caution tag hanging from the unrelated valve FWV–17B. He then noticed that the tag blocked the lights to EF–V12B, which was in direct line from the auxiliary feedwater pumps. Immediately glancing down to its companion EF–V12A, he excitedly flipped the tag out of the way of EF–V12B.

Finally something was making sense. It confirmed what he had hoped was not true, but at least it made sense. So far, the operators had been talking at fairly normal levels even if there was understandable tension. But Faust now yelled to the others: "the 12's are closed!"—immediately interrupting everyone and getting the attention of the room. The other operators came over to Faust's station to witness the opening. Faust grabbed the knobs hard, but then eased up since he expected that he would cause cold water to splash over the extremely hot metals in the dry generators.

He cracked the valves open. The loose parts monitor soon began to "clank" in the control room, telling the operators that cold water was in fact spraying over very hot metal. The emergency feedwater pumps were beginning to come back on line, and the generators would soon begin to fill with water. Eight minutes late, the heat sink for the primary cooling system was being restored.

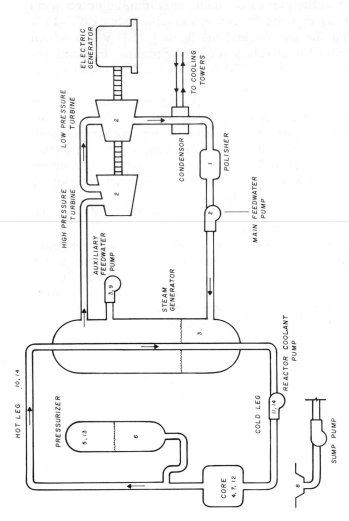

Figure 5. Schematic Diagram of the Accident, March 28, 1979

1. Polisher becomes blocked 4:00:36 A.M. 2. Main feedwater pumps and turbines trip 4:00:37 A.M. 3. Auxiliary feedwater pumps blocked, so generators begin to boil dry 4:00:37 A.M. 4. Reactor trip 4:00:44 A.M. 5. Pressure relief valve sticks open on pressurizer 4:00:50 A.M. 6. Pressurizer fills with water 4:02 to 4:05 A.M. 7. Emergency Core Cooling System turned off 4:03 A.M. 8. Sump pump sends overflow water to auxiliary building 4:08:15 A.M. 9. Auxiliary feedwater pumps unblocked, and begin to refill generators 4:08:30 A.M. 10. Steam voids in hot legs begin to block coolant flow 4:30 to 5:15 A.M. 11. First reactor coolant pumps shut off 5:14 A.M. 12. Core begins to be partially uncovered 5:45 A.M. 13. Pressure relief valve blocked, stopping coolant escape 6:22 A.M. 14. Reactor coolant pump restarted, collapsing voids in hot leg 8:00 P.M.

NOTES

1. "Technical Staff Analysis Report on Control Room Design and Performance" to President's Commission on the Accident at Three Mile Island, October 1979, p. 11.

2. Testimony of Craig Faust (operator) before President's Commission, May 30, 1979, p. 200.

3. Testimony of Gary Miller before President's Commission, May 31, 1979, p. 5.

4. Testimony of William Zewe before U.S. Congress, House, Committee on Interior and Insular Affairs, *Accident at the Three Mile Island Nuclear Powerplant*, Part I, 96th Cong., 1st sess., May 9, 1979, p. 19.

5. Testimony of James Floyd before President's Commission, May 31, 1979, pp. 149-50.

6. Ibid., p. 39.

7. Testimony of Craig Faust before President's Commission, p. 200; and testimony of Craig Faust, *Accident at the Three Mile Island*, Part I, May 11, 1979, p. 120.

8. Testimony of William Zewe, *Accident at the Three Mile Island*, Part I, May 11, 1979, p. 160.

9. Testimony of Edward Frederick, *Accident at the Three Mile Island*, Part I, May 11, 1979, p. 120.

10. U.S. Nuclear Regulatory Commission, "TMI-2 Interim Operational Sequence of Events As of May 8, 1979," May 8, 1979, p. 1 (mimeographed); testimony of Joseph Logan (shift supervisor, Unit II) before President's Commission, May 31, 1979, pp. 142-43.

11. Testimony of William Zewe before President's Commission, May 30, 1979, p. 180.

12. Testimony of Bruce Lundin before President's Commission, May 30, 1979, p. 56. "Technical Staff Analysis Report on Condensate Polisher," p. 5.

13. "Technical Staff Analysis Report on Condensate Polisher," p. 5.

14. Testimony of Darrell Eisenhut of NRC, *Accident at the Three Mile Island*, Part I, May 9, 1979, p. 19.

15. U.S. Nuclear Regulatory Commission, "TMI-2 Interim Operational Sequence," p. 2.

16. Scenario from testimony of Craig Faust, *Accident at the Three Mile Island*, Part I, May 11, 1979, p. 121.

17. Testimony of William Zewe, *Accident at the Three Mile Island*, May 11, 1979, p. 168.

18. Scenario of William Zewe's actions from his testimony, *Accident at the Three Mile Island*, Part I, May 11, 1979, pp. 160-207.

19. The scenario is taken from the testimony of all involved parties before the President's Commission, May 30, 1979, pp. 76-113.

20. Donald Baier, "The Harrisburg Hoax," *Fusion* 2 (May 1979): 10. A list of possibilities is considered in "Technical Staff Analysis Report on Closed Emergency Feedwater Valves" to President's Commission, October, 1979. See also "Investigation into the March 28, 1979 Three Mile Island Accident by Office of

Inspection and Enforcement," NUREG 0600 (Washington, D.C.: Nuclear Regulatory Commission, August 1979), pp. (I-1-35) to (I-1-37).

21. Testimony of Gary Miller before President's Commission, May 31, 1979, p. 41: "Technical Staff Analysis Report on Closed Emergency Feedwater Valves," p. 16.

22. Testimony of Carl Michelson, *Accident at the Three Mile Island*, Part I, May 10, 1979, p. 71.

23. Actions taken from testimony of Craig Faust, *Accident at the Three Mile Island*, Part I, May 11, 1979, pp. 120-57.

24. Actions taken from testimony of William Zewe, *Accident at the Three Mile Island*, Part I, May 11, 1979, pp. 160ff.

25. Actions taken from testimony of Edward Frederick, *Accident at the Three Mile Island*, Part I, May 11, 1979, pp. 124-57.

26. "Technical Staff Analysis Report on Pilot-Operated Relief Valve (PORV) Design and Performance" to the President's Commission, October 1979, p. 7.

27. Testimony of William Zewe before President's Commission, May 30, 1979, p. 126. .

28. Ibid., p. 130.

29. Line printer for TMI-II, 4:01:37 A.M., March 28, 1979.

30. Testimony of Craig Faust before President's Commission, May 30, 1979, p. 168.

31. "Technical Staff Analysis Report on Control Room Design and Performance" to President's Commission, October 1979, p. 3.

32. Zewe does not remember the exact temperature, but remembers that he asked about two and one-half minutes into the event. According to the line printer, the temperature was about 100°F at that time. See also, testimony of William Zewe before President's Commission, May 30, 1979, p. 131.

33. Events in the pressurizer to the point of going solid are taken from U.S. Nuclear Regulatory Commission, "TMI-2 Interim Operational Sequence," pp. 6-7.

34. Testimony of William Zewe, *Accident at the Three Mile Island*, Part I, May 30, 1979, p. 164.

35. Testimony of Edward Frederick, *Accident at the Three Mile Island*, Part I, May 30, 1979, p. 132.

36. Testimony of William Zewe, *Accident at the Three Mile Island*, Part I, May 30, 1979, p. 164.

37. Testimony of Edward Frederick, *Accident at the Three Mile Island*, Part May 11, 1979, p. 137.

38. "Technical Staff Analysis Report on Transport of Radioactivity from the TMI-2 Core to the Environs" to President's Commission, October 1979, p. (4-4).

39. Scenario to end of chapter from testimony of Craig Faust and Edward Frederick, *Accident at the Three Mile Island*, Part I, May 30, 1979, pp. 129-31 and 134.

Chapter 4

Deteriorating into Stability

After the fact, it is now understood that the blocked auxiliary feedwater pumps complicated the problems at TMI, but were not responsible for causing the damage to the core.[1] Nevertheless, the first few minutes after the pumps were unblocked gave the operators the first positive feedback they had received in several minutes.

For one thing, the number of operators available to monitor the equipment was increasing. At the beginning of the problems at 4:00 A.M., there were two operators in the control room, with another immediately available. After two minutes, there were four. Seven minutes into the transient, the number rose to five, as George Kunder moved over from the Unit I control room. Kunder was the shift supervisor and supervisor for technical support in Unit I. When Unit II tripped, it cut off the steam supply helping to heat the water in Unit I. This loss of steam was a fairly serious aggravation since hot water was needed to maintain the integrity of the seals in the cooling system.[2] With both units now shut down, oil burners were needed to maintain water temperature.

Kunder made arrangements with his operators to get the auxiliary heat started. He then called Station Manager Gary Miller at home to tell him that Unit II had tripped. At this point, Kunder knew no details to answer Miller's questions. Therefore, Miller, who had just arisen, continued for the next hour to prepare for his scheduled meeting later that day in New Jersey. Kunder then went to the Unit II control room to offer his services if needed. Entering the control room one minute before the closed feedwater valves were discovered, his services were definitely needed.

But after the feedwater pumps were opened, the parameters of the plant seemed to improve considerably. Within fifteen seconds, there was enough water in the secondary side of the generators for both the hot leg and cold leg temperatures to begin falling rapidly. Also, the pressure in the steam generator began rising toward the 1010 psi that was needed for properly regulated cooling. Faust brought the water level over the next few minutes up to thirty inches, which is the low level limit for normal operation but the prescribed level for emergency operation.[3] Then he throttled back the auxiliary feedwater pumps to avoid overcooling the core. Until more was understood about the problems in the primary coolant loop, Faust's job was to maintain this level. It was an awkward task, but it could be done.

Readings in the primary system were still contradictory, but were beginning to change. Despite the open pressure relief valve, the pressure in the primary system had actually risen a little for the previous two minutes while the operators experimented with the makeup and let-down systems. But within a minute of restoring the generators, the primary temperatures and pressure both began to fall. Two minutes after the blocked valves were opened, even the pressurizer came slightly back on scale, but stayed almost full for the next hour.

The dropping temperatures were due to the restored secondary cooling in the steam generators. However, the dropping pressure was primarily caused by the open pressure relief valve. Unfortunately, with the cooling now taking place, the operators came to the conclusion that the dropping pressure must be caused by the shock to the primary system of pouring cold water into the overheated generators.[4] This theory did not explain why the pressure was so low already, but it explained why it was dropping further. Therefore, the operators did not feel any strong incentive to look elsewhere for an explanation of the declining pressure.

Ten minutes into the transient, a second sump pump automatically started as the overpressurized quench tank leaked increasing amounts of cooling water onto the floor. But all this was happening without the knowledge of the operators. As with the first pump, there was no control room indication that the second sump pump had started.

At this point, Frederick was distracted by the failing makeup pump A. It tripped, and he restarted it. It tripped again, and he restarted it. Then the high sump alarm lit on the annunciator panel to warn that despite the two operating sump pumps, the water was now almost five feet deep on the floor of containment.

Actually, there were two new developments that, had they been noticed at the time, indicated trouble. With two sump pumps removing 280 gallons of water per minute, the water level should not have been rising in containment. Related to this, since it was the cause of the water leakage, at eleven and one-half minutes, the coolant gauge for the quench tank went off scale.

But by now the operators were up against one of the inherent problems of reactor control rooms; in a crisis, the annunciator panels do not present the information in a very useful format. Within the Unit II control room, there are 1600 alarm windows attached to one audible alarm. There is another audible alarm to warn that something is wrong in one of the computer-monitored systems. By now, almost 200 alarm windows were flashing.

Some of the problems were now resolved, and many others were understood by the operators but not yet resolvable. But the only way to quiet the audible alarm or turn off the distracting lights to see what still needed attention was to hit the "annunciator acknowledge" button.

This is a useful button under normal circumstances, allowing an operator to immediately stop any alarm on the panel and clear the board for new signals. But hitting the "annunciator acknowledge" button would mean that all continuing problems would become solid rather than flashing lights and that all previous but now resolved problems would disappear from the board entirely. Frederick decided early in the transient that the circumstances were confusing enough that it might be useful later to go back and retrace what had malfunctioned. The only way to do that was to refuse to acknowledge the alarms. He asked the other operators not to silence the alarms, and they did not. Now, ten to fifteen minutes into the event, he was studying 200 flashing lights while restarting the failing makeup pump A two more times. Somewhere buried in those lights was the high sump alarm.

Besides the annunciators, there is one other way to call up information on a plant malfunction. All major shifts, including annunciator alarms, are automatically printed out on an IBM typewriter computer terminal in the control room. This is an impressive machine that can type information from the computer at a few hundred words per minute. However, during the opening moments of the transient, it was receiving ten to fifteen messages per second.

It simply backlogged these and printed them out in the order received, two to three hours later.[5]

The sump problem continued to worsen. For several minutes, the quench tank had been venting excess pressure into containment and, therefore, onto the floor. However, the stuck pressure relief valve was adding more pressure than the vent could relieve. A few seconds before fifteen minutes into the transient, the quench tank pressure reached 192 psi. To avoid damaging the tank, the rupture disk placed on the tank for this purpose failed as designed and dumped over 7000 new gallons of core-cooling water into containment.

The only notice to the operators in the control room was the high sump alarm still flashing but lost among the lights. There was also a level indicator, but there was no reason to check it since the high sump alarm had not been noticed. The water level on the floor of containment was now approaching six feet.

About sixteen minutes into the event, Sheimann and Frederick were manipulating primary coolant controls to try to get the pressurizer level down. The level indicator seemed to be teasing them by fluctuating a little below completely solid. As they worked, the flow through the primary system was gradually becoming more strained as the steam voids interfered, but this development was not yet obvious to the operators. Faust was manually operating the auxiliary feedwater pumps and the throttle valves to maintain a thirty inch level in the generators, and this was a full-time task. Kunder was trying to review the developments so far to make sure that no appropriate steps had been forgotten.[6] Zewe decided to try to take the pressure off Faust by restarting the secondary coolant loop. For this, he went to the turbine building.

He actually managed to get a condensate pump started. However, without realizing that he had stumbled on the problem that had started the entire transient, he was unable to get a flow of water through the polishers. He tried to bypass the polishing system, but the flooded controls would not respond. Not understanding the problem, but noticing that Faust was able to control the auxiliary feedwater level satisfactorily, Zewe moved on to look at other systems back in the control room.

The atmosphere in the control room changed noticeably in this period, around fifteen to twenty minutes into the transient. In the first few minutes, there had been so many alarms that the operators had been forced to act quickly and by their training to counteract the crisis. There had been too many things to do and not enough—if any—time to think things through.

But the restored secondary feedwater had changed the sensitivity of the unit. The operators were to face crises over the next few days that made the opening minutes seem minor by comparison. Crises now developed over periods of minutes or hours, not fractions of a second. There was time to think. As a result, the operators began to discuss the situation at greater length and to turn levers less frequently.

One situation that needed thinking through occurred about twenty minutes into the event, when the operators received some unsettling news. The neutron countrate, one method of measuring fission in the reactor, was beginning to rise. While discussing the possible reasons for this, the laboratory called with the results on a sample of core-cooling water that they had drawn through their normal sampling lines shortly after the reactor trip. The operators expected the boron concentration to be elevated due to boron in the makeup water tanks. Instead, the boron count was dropping rapidly. Also, the radioactivity in the core water was rising.

As if they did not have enough problems, the core was beginning to reheat. They were losing the neutron-absorbing boron, and fission was slowly restarting.

For the moment, these readings were more baffling than actually threatening. The control rods were still inserted, and a runaway core was out of the question. Nevertheless, excess boron in the makeup water was one of the safety features of the reactor, and the boron was disappearing. The operators began to reach the conclusion that there must be a substantial breach of the primary coolant loop at some point. Their next task was to decide where it must be.

Almost twenty-five minutes into the transient, Zewe asked for a computer reading on the thermocouple outside the pressure relief valve. He later said that he asked for the reading as part of "the normal alarm status review."[7] He also remembered the reading as 220° F.[8] Nineteen days later, he remembered the reading as 228° F.[9] In fact, it was 285.4° F.[10]

But any of these readings could be considered ambiguous. Various experts in testimony before the President's Commission and Congress have disagreed on what reading a fully open pressure relief valve would cause, but none guessed below the low 300s° F. On the other hand, a closed valve normally reads below 200° F but does occasionally exceed that in normal leakage. A reading in the intermediate range required the operators to do some interpretation.

The thermocouple was strapped to the outside of the exit pipe and measured the temperature of the pipe itself. Twenty-five minutes into the transient, it was normal to expect the pipe to be holding some heat from the earlier release. The reading may have been a little high for these circumstances, although the operators were not sure that it was.[11] If Zewe actually understood the reading to be in the 220s° F at the time, the reading was not too high at all. At any rate, it made more sense to the operators that the earlier release had now cooled to this level than that an open valve was releasing only this much heat.

Within a few minutes, the transient was thirty minutes old. The primary pressure was less than 1200 psi, not much more than the normal pressure in a boiling water reactor. The core-cooling water temperature was reduced, but still exceeded 500° F. The pressurizer was virtually full. Two makeup pumps, 1C and the troublesome 1A, were operating but throttled back.

The secondary cooling system was still blocked, but Faust was holding the generator levels at thirty inches as prescribed. However, this was becoming more difficult, and his control of generator B began to falter. Finally, at thirty-six minutes into the event, he shut down the feedwater pump for generator B and used water diverted from the other two pumps.

About this time, the control room received a message that enabled them to explain in their own minds many of the mysteries that had been reflected in the confusing monitor readings. An auxiliary operator near the three waste disposal tanks in the auxiliary building had been watching his water levels rise and had been intrigued by the rising water level on his remote monitor for containment. It showed the water in containment to be in excess of six feet deep and therefore off his scale.

He called the control room to alert them that he had these readings. Zewe checked his instruments, noting the high sump alarm and reading that their indicator was also off scale. He noticed that the containment building pressure was at 2 psi, indicating that something was leaking into the containment atmosphere. With two pumps operating, but continued high water and pressure, Zewe decided that he had verified the suspected leak. If the building pressure increased to 4 psi, automatic controls would isolate containment from all other buildings. The leak was not that severe. However, since they had already received an alarm for elevated (but not dangerously high) radiation, Zewe shut off the two sump pumps to stop the flow to the auxiliary building until the source of the water could be verified.

Having discovered the leak, the operators still did not know where it was. However, they had been encountering some difficulty in maintaining the level and pressure in generator B. Were they to assume for a moment that generator B was leaking, most likely through a feed line break, that would explain the water on the floor of containment.[12] It would also account for the elevated pressure in containment. If generator B was also forcing secondary water into the primary system through internal ruptures, that would explain the declining pressure that had been noticed in the secondary side of generator B. It would also explain the disappearing boron, since the entering secondary water would be diluting it.[13] Ironically, a leak would normally force water into the secondary loop, but primary pressure was now below secondary pressure.

They tried a test by isolating generator B for a short period. Primary loop pressures dropped, leading the operators to believe that the flooding of the primary system by the secondary system had been cut off.[14] In fact, there was even a good explanation for the failure that they now suspected, since the generators had just been through the double shock of drying out and then getting doused by cold water. The theory was so useful in explaining the last few minutes that it had to be true.

Long after the accident, there is still some evidence that generator B may be leaking. But it was not the explanation for the crisis on March 28. The water and building pressure were coming from the open pressure relief valve. Still, the leaking generator theory was so helpful in tying several confusing readings together that the operators clung to it. Hours later, after the pressure relief valve had been discovered and closed, they went so far as to isolate the generator. But for now, the operators cut the second of three auxiliary feedwater pumps and kept the generator on line. The situation did not seem to warrant emergency actions so long as the operators felt that they knew what was happening.

For perhaps thirty minutes, the plant continued to operate in this mode. Core temperatures were in the low 500s° F, and the pressure hovered around 1100 psi. Water was passing through the pressurizer onto the floor of containment. With the makeup pumps running, the core remained covered. But steam voids continued to build in the primary loop, and the four reactor coolant pumps were now clearly straining their 10,000 horsepower motors to move the water. The operators were at a loss to explain why.

After 5:A.M., the reactor coolant pump problem became very serious. The difficulty of pushing water through the steam voids was becoming more than the pumps could handle, and they were approaching a condition known as cavitation. Reminiscent of the climatic scene in the movie *China Syndrome*, the pumps began to vibrate violently on their supports. The flow of water through the pumps was oscillating and threatening to stall. The operators had been trained that there was a specific level of vibration beyond which the seals in the pumps could be damaged. By 5:13 A.M., the pumps on the B side of the core were at three times this level.

While the operators were deciding what to do about the pump problem, Station Manager Gary Miller called the plant to talk to George Kunder in the control room. Miller was almost ready to leave for his meeting that day in New Jersey. Before leaving, he wanted to verify that the reactor was under control and to find out what had caused the reactor trip.[15] When Kunder outlined the unexpected and unexplainable difficulties, Miller asked Kunder to set up a conference call with himself, Met Ed Vice-President for Generation John Herbein, and someone from Babcock and Wilcox. He then went back to preparing for the trip and waiting for the call.[16]

Back in the control room, shortly after 5:14 A.M., the operators agreed that the reactor pumps on the B side of the core would have to be turned off or they would damage their seals. Faust took the controls for the two pumps and shut them off. With the heat from the core and the cooling in the generator, this water should still have flowed through a process called natural circulation. However, the operators did not believe that it would, since they felt the generator was damaged. Accordingly, they did not even raise the water level in the secondary side of the generator to the amount needed for successful natural circulation.

The operators were instead depending on the system to continue cooling the core through just the A side of the primary loop. In normal circumstances, that can be done without difficulty.[17]

As an added aggravation, the line printer also broke down. While continuously typing for well over an hour, the typewriter had been using a large supply of paper. At 5:14 A.M., some of the paper fed in unevenly, and the paper feed mechanism jammed. As a result, the typewriter could no longer operate.

Information in the computer was still accessible through a video display screen, and the typewriter was hours behind anyway. However, the typewriter had been the best method for reconstructing the early moments of the event so as to try to explain the current readings. An attempt was made to free the jam, but it was severe. The typewriter stayed out of operation for over one and one-half hours, and some of the information fed to the typewriter during this period overloaded the memory and was permanently lost.

Shutting the B reactor coolant pumps down had not improved the situation in the primary coolant loop. At 5:21 A.M., Zewe again asked for the readout on the pressure relief valve. The reading was 283° F. After a short discussion, the operators again decided that the valve must be closed and that the leak must be coming instead from generator B.

But possible problems in generator B did not explain the problems with the reactor coolant pumps. Once the reactor coolant pumps began to oscillate, evidence that there were voids in the system could be considered overwhelming.[18] However, relying heavily on the fact that the high radiation alarms were not active, the operators concluded that the system must still be virtually solid.[19] Therefore, they continued to work under that assumption.

Over the twenty-five minutes following the shutdown of the B pumps, the flow in loop A became increasingly strained. The pumps began to oscillate and vibrate until they reached the same dangerous readings that had been observed in the B pumps.

It is now believed that these pumps can be run under any circumstances without damaging them.[20] Since this incident, the NRC has instructed operators to run the reactor coolant pumps during an emergency to the point of breaking them.[21] But the operators had specific guidelines on the safe levels of operation of the pumps, and they felt that they had no choice. They might need these pumps later and could not use them if they had been damaged. Therefore, at 5:41 A.M., Faust and Frederick jointly took the controls and closed off the last two reactor coolant pumps.

Because the pressure relief valve had been open so long, serious damage to the reactor was probably already inevitable. However, when the operators shut off the reactor coolant pumps, they had forced the crisis to a head. Without knowing it, they had uncovered part of the core.

In their final minutes, the struggling reactor coolant pumps had managed to keep the system deceptively stable. Primarily, the pumps had been pulling what water they could through the generators and forcing it into the core vessel. Beyond this, only small amounts of water made it through the almost entirely voided hot legs. Therefore, the water was trapped in the core, and there was more water in the core and less in the generators than would naturally have been the case.

With the pumps stopped, the water slowly sloshed out of the core and back into the cold legs at the exits of the generators.[22] An inadequate amount of water was seeking its own level, and it slowly uncovered about four feet of the top of the fuel assemblies. This process took about ten to fifteen minutes to develop since the water had to flow backwards through pumps and steam voids. During this period, the operators were not aware that this was taking place.

Nevertheless, there was some degree of concern about the confusing conditions that seemed to be getting continually worse. About this time, Richard Dubiel reached the control room. He was the Director of Chemistry and Health Physics for both units and in charge of monitoring radiation in the plant. Kunder had called him earlier, and he now arrived to take over the operation of the lab and radiation-monitoring equipment.

Because there was high water on the floor of containment from an unspecified source, Kunder asked Dubiel to take a radiation and chemistry reading in containment. Unfortunately, the normal monitor in containment was now under water.[23] Therefore, someone had to enter containment and "grab" a sample of the water. The technician who did this managed to get his hands wet from the water while drawing a sample. However, there was no readable contamination on his hands, and Dubiel reported to Kunder that the water in containment appeared to be from steam, but had a low level of radioactivity.[24]

Kunder also wanted a reading of the core-cooling water, not realizing that it was the same as the water in containment. For this, there is a sampling line that runs from the primary system for several hundred feet over to the "hot lab" in Unit I. The samples drawn in the hot lab normally reflect the conditions in the primary loop forty minutes previously.[25] It is an awkward system, but is designed to protect the technicians from possible contamination during an emergency. Dubiel reported back to Kunder that there was a low but elevated level of radiation and that the boron count was reduced. This merely confirmed earlier readings.

Not realizing there were voids in the coolant loop, the operators assumed that cooling could be achieved through natural circulation. Accordingly, Faust began increasing the water level in generator A from thirty inches to the twenty-one feet needed for natural circulation cooling. This took eighteen minutes, but the delay was not important, since the primary water was barely circulating to be cooled anyway. During this period, the hot leg temperatures began to rise, as the limited circulation that had been forced by the reactor coolant pumps was now almost completely stopped.

It was now almost two hours into the transient and still thirty minutes before the open pressure relief valve was discovered. From hindsight, it is easy to argue that the low pressure and elevated temperature were ample evidence of both a leak in the system and voiding in the core and hot legs. In fact, these possibilities continued to be discussed in the control room. The question of voiding in particular was now interesting to the operators because it was about the only reasonable explanation for the strain on the reactor coolant pumps. However, the operators were convinced that voiding would cause the core to overheat to at least the extent that radiation alarms would be set off.[26] Because these discussions were happening while the excess water above the core was still flowing away, the core had not yet been damaged enough to activate the alarms. Before the alarms sounded, new actors had entered the scene.

Outside the plant, word had been spreading among Met Ed officials that something was wrong at the plant. The first to arrive was Joseph Logan, who entered to take nominal charge of the control room at 5:45 A.M. As soon as he saw the pumps stopped, the low pressure, and the full pressurizer, he went to the supervisor's office to find out from Kunder what was going on.[27] Zewe was busy at the control panels, and Logan did not want to disturb him.

Over the next hour, Logan was the supervisor in charge of the control room. He spent this time speaking to Kunder, Zewe, and others who entered to help. Several actions were taken during this period after common discussions in the control room. However, Logan never actually ordered any specific moves. The President's Commission later found this rather odd and questioned him with some degree of abruptness.[28] Obviously, it would have been refreshing to have a dynamic personality walk in at this point who understood the nature of the problem and could give the appropriate orders and leadership.

Unfortunately, no one in the control room understood the situation well enough to give such orders. Rather, other than those directly operating the control panels, the people in the room spent their time analyzing and discussing the system parameters. Furthermore, this group of discussants soon began to grow rapidly.

Shortly after 6:00 A.M., a conference call was finally arranged among Kunder in the control room (with others contributing information), Miller at home, Met Ed Vice-President Jack Herbein in Philadelphia, Babcock and Wilcox on-site representative Leland Rogers, and George Cutter, a technical assistant from Unit II. They discussed the history and current readings of the transient. They decided that the ruptured disk on the quench tank must have come from the original pressure release, since they assumed that the pressure relief valve was now closed. During this call, the core was in the process of uncovering. However, steam flashing from the cooling water was still enough to keep the zirconium coating on the fuel pellets from rupturing, and there was no radiation alarm. Therefore, they did not realize that the core was in imminent danger.

It was soon clear that the phone conversation was not going to solve this growing mystery. Therefore, Jack Herbein asked Gary Miller to cancel his trip to New Jersey and to go instead to the plant. He also asked that some of the extra personnel who had earlier been called into Unit I to help with its steam problems now be transferred to Unit II as needed. The conversation ended, and Gary Miller began calling the others who were to share a ride that day to New Jersey. At the plant, people began to come over from Unit I.

By 6:12 A.M., steam generator A was at its natural circulation level. However, the temperature in the hot leg was climbing and was almost off scale at 620°F. Temperatures in the B hot leg were even higher and were off scale. The reactor coolant pumps were still turned off.

As new people entered the room, the process of bringing them up to date allowed the operators already there to reassess what had transpired. One of the people to enter during this period was Brian Mehler, who would have been the shift supervisor on the next shift. In analyzing the information available, he asked Zewe to check the pressure relief valve again. Zewe of course had thought of this and had checked it twice earlier. However, he called the temperature up on the computer terminal again at 6:18 A.M. and received a reading of 228.7°F.[29]

Zewe was later consistent in contending that he remembered no reading during that morning above 232°F. It is possible that he understood that to be the maximum on the morning of March 28, or it is possible that he later forgot the readings in the 280s°F. At any rate, it struck both Zewe and Mehler as suspicious that the residual heat in the pipe was still this high over two hours after the release. Therefore, they agreed that they should try isolating the valve by closing the block valve downstream.[30] The block valve stops any flow in the pipe that gets by the pressure relief valve.

At 6:22 A.M., almost two and one-half hours after the pressure relief valve had jammed open, Sheimann took the control for the block valve and isolated the flow escaping through the pressure relief valve. The hole in the primary coolant loop was now plugged. Far too late, the cause of the nation's most serious threat of a meltdown had been corrected.

For the second time since the transient had begun, the readings in the control room almost unanimously began to improve. In fact, the improvements seemed more substantial now than they had after the auxiliary feedwater valves had been opened. Ironically, this was happening as zirconium coating on the fuel pellets was beginning to crack.

The operators quickly recognized that the pressure relief valve had in fact been open. The pressure in the primary coolant loop, which had dipped below 700 psi, began to rise dramatically and continued for the next thirty minutes until reaching almost normal operating levels. With the escape now blocked at the top of the pressurizer, the water level in the pressurizer dropped to high, but acceptable, levels. The containment building, which had been slightly pressurized to about 2.2 psi, rapidly dropped back to almost atmospheric levels now that the escape of steam had stopped.[31]

Unfortunately, the operators were not able to check the core temperatures. With the other improving conditions, it would be normal to expect these readings to be dropping. However, the thermocouples in the core were intended to monitor normal heat distribution and were not designed for emergency conditions. The readings had already climbed off scale and remained there. The only available reading was the "average temperature" in the primary loop, which was indicating a decrease because of a flaw in its calculating technique.[32]

There was no instrumentation capable of taking the necessary readings, and it is now impossible to calculate exactly what was occurring in the core at this point. However, with the declining water level, it is virtually certain that the temperatures were a few thousand degrees and climbing.[33] In addition, by 6:30 A.M., Zewe picked up the first reading that led him to suspect that the core itself might be in trouble. The neutron countrate began to climb again. This normally means that the level of core activity (fission) is increasing, although at this point it probably meant that the water to slow the neutrons was disappearing.[34] At any rate, it should not have been happening unless something was seriously wrong in the core.

For several more minutes, the operators monitored the conditions under the assumption that the plant was returning to normal. At 6:30 A.M., with pressures increasing, Faust returned to generator B and began to raise its secondary water to natural circulation levels. Frederick, monitoring the primary loop, watched the rising pressure levels with relief. However, the pressurizer water level was also climbing from the low of about 300 inches that it had reached shortly after the pressure relief valve had closed.[35] Soon it was off scale again at 400 inches. Frederick again decided that the system must be solid, and at 6:45 A.M., he cut one of the two operating makeup pumps.

At 6:45 A.M., just as the end of the crisis seemed near, time finally caught up with the operators. All pretense that the reactor might still be returned unharmed to a stable condition collapsed. For the previous hour and five minutes, all the reactor coolant pumps had been stopped. For most of this period, the pressure relief valve had also been open. Because of this combination, the water around the core had been flashing to steam or flowing back into the cold legs. About four feet of the top of the fuel assemblies were uncovered. At first the zirconium coating on the fuel pellets began to oxidize. Then the coating began to crack, exposing the uranium dioxide inside. This was not immediately obvious because the radiation monitors were somewhat removed from the core to protect them from the heat. By 6:45 A.M., some of the zirconium was probably beginning to melt.

Dubiel in the hot lab in Unit I had been drawing samples since arriving an hour before. He had been picking up radioactivity readings in the core water of a scant few millirems per hour. Suddenly at 6:45 A.M. the readings jumped to several hundred millirems per hour. Dubiel called Kunder in the control room to report the new development. As he was doing this, several radiation monitors in the control room started flashing.

Ever since the reactor coolant pumps had started to cavitate, the operators had leaned more heavily on the lack of radiation alarms as evidence that the plant was still solid. Suddenly, the radiation was intense. It was now clear to the operators that the primary loop was voiding badly.[36] With this level of radiation, it could also be assumed that the fuel itself had been damaged.[37] While they still felt certain that they had tools to fight it, they had passed the first step in a potential meltdown.

Meltdowns are one of the more speculative and more controversial aspects of nuclear technology. On paper, their potential for destruction is overwhelming. According to the best available theories, the residual heat left in a core with a long-term loss of coolant would be enough to turn the zirconium coating on the uranium fuel pellets into molten metal. Once this happens, the uranium would condense into a large superheated mass that would be extremely radioactive. Given the heat of the mass, it would also melt anything that tried to contain it. Within a matter of hours, it would melt through the core vessel. Within days, it would disintegrate the concrete under the containment building and dig into the earth. Theoretically, it could melt its way into the ground indefinitely in what is now popularly known as the China Syndrome. More likely, it would hit the water table and cool and disperse while flashing the water to steam.[38]

Under the best of circumstances, a full meltdown would be catastrophic. However, we have never experienced anything close to a meltdown on a scale approaching that of a commercial reactor. We have the technology to create such a catastrophe if we should ever have the bizarre desire. However, the circumstances inside a commercial reactor are complicated enough that it is not clear what kinds of believable accidents could lead to one.

Certainly, a fully open let-down system and artificially shut makeup system would drain the core and eventually cause a meltdown. However, in the case of TMI, there was only an imbalance in these systems. More water was being released through the let-down system than added through the make-up system. However, there was still some water in the core vessel that was partially cooling the core by flashing to steam.

There is no way to know for sure whether a meltdown would ever have occurred at TMI so long as some new water was reaching the core. It was two days before a total loss of water seemed possible. However, it is clear that during this initial crisis, the core was in danger of nothing more than substantial fuel damage.[39]

Rather, the most significant immediate danger to the plant on the first day was not in the core but in the release of radiation. On the issue of containing radiation, the plant was poorly prepared. Should there ever be a radioactive release from the primary coolant loop, it is supposed to be confined by the containment building. This building has concrete walls three to four feet thick to stop all but minute amounts of radiation. There is a double lock door system on containment buildings to assure that at least one door to the outside is always closed.

But containment buildings are actually full of holes. Pipes and electrical cables run through the walls at numerous places to provide such services as control for the reactor, drainage, and outlets for steam to go to the turbines. More importantly, the makeup and let-down systems to regulate the amount of water in the primary loop have their components outside containment.

Under normal operating conditions, containment is nothing more than a potentiality. But once anything foreign is introduced into the building's atmosphere, the building is automatically isolated. The mechanism for doing this at Three Mile Island was an automatic control to seal the containment building when the overpressure reached 4 psi. That setting was high enough to prevent the mechanism from accidentally engaging. However, it was supposed to be low enough to engage when any releases occurred in the building.

By hindsight, the automatic sealing mechanism was set too high. For almost two and one-half hours the pressure relief valve vented the core-cooling water into the building and raised the building overpressure to only 2.2 psi. As the radioactivity soared in the primary loop, the sump pumps were off, but radioactivity could still be released most heavily through the let-down valves into the makeup tanks in the auxiliary building and through the quench tank waste gas vent into vent headers in the auxiliary building.[40] In fact, it is now believed that gas buildup in the quench tank damaged the maze of routing pipes and valves called the vent header in the auxiliary building during this period, leading to leakage problems later in the accident.[41]

By 6:50 A.M., substantial amounts of radiation were setting off alarms in both containment and the "hot shop" in the auxiliary building. According to the emergency plans devised at the time TMI was licensed, there are three levels of emergency—local emergency, site emergency, and general emergency. When radiation was recorded in a second building (the auxiliary building), TMI was by definition into a site emergency.

Within the control room, strong actions were now attempted to regain control of the clearly deteriorating reactor.[42] Frederick and Sheimann attempted to start reactor coolant pump 2A, but the voids partially blocked the pump mechanism and the pump refused to start. They tried pump 1B, but it was also blocked and refused to start. Finally, pump 2B resisted but did begin to operate and struggled for eighteen minutes before it had to be stopped again. Also, to maintain all the water they could in the primary loop, the operators isolated generator B, which they had suspected of leaking for the past hour.

The pressure was responding well in both the primary and the secondary loops. On the secondary side, steam generator A threatened to overpressurize. At 7:04 A.M., it was bypassed to the noisy atmospheric dump valves, alerting those in nearby Goldsboro across the river that something was amiss. On the primary side, the pressure reached 2130 psi, and the operators began to manually operate the block valve on the pressure relief valve. These measures kept the pressures stable, but could not force the water to flow.

About 7:15 A.M. Gary Miller arrived to take charge of the control room. Because of his earlier phone conversations, he knew roughly what to expect. However, what caught him most by surprise was that radiation monitors were now flashing at several stations in the plant and that the radiation was rapidly growing in intensity.[43]

Miller began selecting individuals from the growing crowd and assigned them to individual controls. Two particular assignments took substantial numbers of personnel and had high priorities. One was radiation monitoring. When a site emergency is declared, the utility is required by its plan to send teams to monitor for radiation on and off site. Miller assembled the teams and sent them on their way.

The second task was notification. When the site emergency was declared, the utility was obligated by its plans to notify several local, state, and federal authorities. This calling was already underway. Miller assured that there were sufficient personnel to make calls and to provide support services.

Within minutes, at 7:24 A.M., the calls had to be started again. Radiation was now extreme in several parts of the plant and had passed the guideline of 8 rems per hour in the dome of containment. Using the official guidelines, Miller declared that the plant had shifted to the highest phase of its emergency plans. TMI had reached a state of general emergency.

NOTES

1. Testimony of Carl Michelson before U.S. Congress, House, Committee on Interior and Insular Affairs, *Accident at the Three Mile Island*, Part I, 96th Cong., 1st sess., May 10, 1979, p. 52. "Technical Staff Analysis Report on Alternative Event Sequences" to President's Commission on the Accident at Three Mile Island, October 1979, p. 18.

2. Testimony of Gary Miller, *Accident at the Three Mile Island*, Part I, May 11, 1979, p. 165.

3. "Staff Report on the Generic Assessment of Feedwater Transients in Pressurized Water Reactors Designed by the Babcock and Wilcox Company," NUREG 0560 (Washington, D.C.: Nuclear Regulatory Commission, May 1979), p. (2-8).

4. Testimony of Edward Frederick, *Accident at The Three Mile Island*, Part I, May 11, 1979, p. 133.

5. Ibid., p. 137.

6. Testimony of William Zewe, who mistakenly identified him as Ken Byron, *Accident at The Three Mile Island*, Part I, May 11, 1979, p. 167.

7. Testimony of William Zewe, *Accident at the Three Mile Island*, Part I, p. 170.

8. Ibid., p. 169.

9. Testimony of William Zewe, President's Commission, May 30, 1979, p. 126.

10. Line printer, TMI Unit II, March 28, 1979, 4:25:35 A.M.

11. Testimony of Gary Miller, *Accident at the Three Mile Island*, Part I, May 11, 1979, p. 170.

12. Testimony of Zewe, Faust, and Frederick, President's Commission, May 30, 1979, pp. 141-46.

13. Testimony of William Zewe, President's Commission, May 30, 1979, p. 209.

14. Testimony of Edward Frederick, *Accident at the Three Mile Island,* Part I, May 31, 1979, p. 141.

15. Testimony of Gary Miller, President's Commission, May 31, 1979, pp. 5-6.

16. Testimony of Gary Miller, *Accident at the Three Mile Island,* Part I, May 11, 1979, p. 165.

17. Testimony of Edward Frederick, *Accident at the Three Mile Island,* Part I, May 11, 1979, p. 140.

18. Testimony of Carl Michelson, *Accident at the Three Mile Island,* Part I, May 10, 1979, p. 61.

19. Testimony of Edward Frederick, *Accident at the Three Mile Island,* Part I, May 11, 1979, pp. 140-41.

20. Testimony of Craig Faust, President's Commission, May 30, 1979, p. 193.

21. Ibid.

22. Testimony of Carl Michelson, *Accident at the Three Mile Island,* Part I, May 10, 1979, p. 57.

23. Testimony of Richard Dubiel, President's Commission, May 31, 1979, p. 66.

24. Ibid., p. 66.

25. Testimony of Gary Miller, *Accident at the Three Mile Island,* Part I, May 11, 1979, p. 179.

26. Testimony of Edward Frederick, *Accident at the Three Mile Island,* Part I, May 11, 1979, pp. 140-41.

27. Logan's activities from testimony of Joseph Logan, President's Commission, May 31, 1979, pp. 125-46.

28. See particularly President's Commission, May 30, 1979, p. 130.

29. Line printer, TMI-2, March 28, 1979, 6:18:30 A.M.

30. Closing of the valve recounted in testimony of William Zewe, President's Commission, May 30, 1979, p. 119.

31. Testimony of Edward Frederick, *Accident at the Three Mile Island,* Part I, May 11, 1979, p. 141.

32. This reading averaged the hot leg and cold leg temperatures. The hot leg was above scale and rising; the cold leg was on scale and falling. The computer calculated the hot leg at top of scale, and not at its actual value, resulting in a false declining reading. Ibid.

33. "Technical Staff Analysis Report on Thermal Hydraulics" to President's Commission, October 1979, p. 8.

34. Testimony of William Zewe and Gary Miller, *Accident at the Three Mile Island,* Part I, May 11, 1979, p. 182.

35. Nuclear Regulatory Commission, *Staff Report on the Generic Assessment,* p. (1-8).

36. Testimony of Gary Miller, *Accident at the Three Mile Island,* Part I, May 11, 1979, p. 179.

37. Testimony of Herman Dieckamp, President's Commission, May 30, 1979, p. 12.

38. David Bodansky and Fred H. Schmidt, "Safety Aspects of Nuclear Energy," in *The Nuclear Power Controversy*, ed. by Arthur W. Murphy (Englewood Cliffs, N.J.: Prentice-Hall, 1976), pp. 33-35.

39. Testimony of Carl Michelson, *Accident at the Three Mile Island*, Part I, May 10, 1979, p. 62.

40. Testimony of James Floyd, President's Commission, May 31, 1979, p. 165; "Investigation into the March 28, 1979 Three Mile Island Accident by Office of Inspection and Enforcement," NUREG 0600 (Washington, D.C.: Nuclear Regulatory Commission, August 1979), p. 15.

41. "Technical Staff Analysis Report on Transport of Radioactivity From the TMI-II Core to the Environs" to President's Commission, October 1979, p. (1-2).

42. Operator actions to end of chapter from "Preliminary Sequence of Events TMI 2 Accident of March 28, 1979," in Nuclear Regulatory Commission, "Staff Report on the Generic Assessment," pp. 8-9.

43. Testimony of Gary Miller, *Accident at the Three Mile Island*, Part I, May 11, 1979, p. 183.

Chapter 5

The NRC Responds

By comparison to the modern-day floodgate of investigations opened by the Three Mile Island crisis, federal regulation of the nuclear industry used to be simple to understand. The Nuclear Regulatory Commission had almost exclusive jurisdiction over nuclear reactor operations in the United States. It licensed each plant, approved the proposed operating procedures, required changes in the procedures and plant structure whenever this seemed appropriate, required the utilities to report when they had violated any of these agreements, and then adjudicated these suspected violations. As a result, while the federal involvement in the TMI crisis over the first few days was to grow into a circus of uncoordinated overlapping efforts, the first and the strongest federal response to TMI clearly belonged to the NRC.

The NRC, formerly the Atomic Energy Commission (AEC), was created by the Atomic Energy Act of 1946. From its enabling legislation, the AEC was given two major objectives relating to the possible development of commercial nuclear power in the United States. First, the AEC was to promote this development, helping make it technologically, economically, and legally attractive. Second, it was to assure that any such development was responsive to the needs of public safety.

There were no commercial nuclear power plants operating or planned in 1946. Instead, for the first several years the new Department of Defense was the AEC's only customer. As a result, the early emphasis of the AEC had to be on promoting nuclear energy. Unfortunately, that was such a struggle that by the time commercial nuclear energy became a reality, the AEC's promotion activities had turned into something of an obsession.

Among the strongest of the early cynics in this field were the utility companies themselves. Given the uncertainties of the new technology, it was not immediately clear to the companies that a nuclear-powered generating plant could be made to operate safely. If it could, it was equally uncertain that it would be economically competitive with coal- or oil-burning plants. Through most of the 1950s the AEC encouraged, cajoled, and finally threatened the utilities to begin building nuclear plants. To a company, they resisted. But three important events finally broke the nuclear ice in 1957, even if only weakly.

First, the insurance problem was solved. As reillustrated by the Three Mile Island experience, we still do not know how effective back-up safety systems are or how much damage a complete meltdown can cause. But the potential damages from a complete meltdown could almost certainly swamp the treasury of any insurance conglomerate. Any claims not covered by insurance could similarly bankrupt the largest utility companies. As a result, no utility could get insurance before 1957 or seriously plan to begin construction of a nuclear plant without it.

This difficulty was solved by the rather unorthodox Price-Anderson Act of 1957. By the provisions of this law, which has been renewed and is still in effect, damage settlements against a company for a nuclear accident cannot exceed a total of $560 million. No matter how catastrophic the accident might be, no more than $560 million can be collected as a total for everyone as a result of any one accident.

Furthermore, $500 million of the coverage is provided free to the utility by the federal government. The utility must then supplement this with $60 million in commercial insurance. After Three Mile Island, the assumptions behind the Price-Anderson Act may seem arbitrary and financially hopelessly out of date. However, they provided and still provide a massive financial subsidy to underwrite the insurance costs in developing nuclear energy.

The second development in 1957 was an AEC threat to the utilities. For ten years, the AEC applied increasing pressure on the utilities to adopt the use of nuclear energy. Over President Eisenhower's objections, the AEC sponsored four separate promotional campaigns to entice utilities, including such incentives as free research and development and a free first load of fuel. The utilities still balked. Finally, in 1957, the AEC announced that if the utilities did not submit proposals soon, the AEC would build its own reactors and begin competing with the utilities for the sale of electricity.

Third, in 1957 the first pilot project reactor went commercial. Ironically, considering the events twenty-two years later, it was in Pennsylvania, at Shippingport. Within a few years, several other pilot plants were operating, including a helium-cooled experimental plant just down the Susquehanna River from the Three Mile Island site at Peach Bottom, Pennsylvania.

While the AEC tried to make nuclear energy as attractive as possible, the utilities in the 1960s were living in an economy that also subsidized cheap coal and oil. But with new pollution control requirements, continued AEC pressure, and finally OPEC, the nuclear option to oil and coal finally began to look much better to the utilities in the late 1960s and early 1970s. Applications for nuclear plants skyrocketed.

In the midst of this new proliferation of interest, issues such as safety systems and the storage of nuclear wastes took on new urgency. But the AEC was tainted with the reputation of twenty years of often brutal pressure to commercialize reactors. Their reputation as a protector of public safety was therefore hampered by their unabashed enthusiasm for the technology. As a result, there was considerable sentiment from both within and without the AEC for a change.

The conflict was finally resolved in 1974, when the regulatory side of the AEC was renamed the Nuclear Regulatory Commission. The promotion and development side joined the newly created Energy Research and Development Administration, which has since been reorganized twice and is now scattered within the Department of Energy.

As the regulatory side, the new NRC was now administratively free to strike a significant blow for reactor safety. It maintained the powers of licensing and regulation so that each of these decisions could be made by people specifically employed to protect the public safety. In this regard, the NRC also created a notification program that gave it the potential to monitor any emergencies that might arise. But its first test in monitoring an accident came perhaps too quickly, and it failed miserably.

On the north shore of Wheeler Lake, a part of the Tennessee River near Decatur, Alabama, lies the Brown's Ferry nuclear powerplant. By sheer size, it is now an impressive sight. But on March 22, 1975, it was still being built. Unit I was in operation and Unit II had gone commercial just three weeks earlier. Unit III was under construction.

As controls were added in the control room for Unit III, the wires for those controls ran through holes in the control room floor to a concrete room below called the cable spreader room. Here the amazing array of wires running from almost all of the controls for all three units were routed into cable trays and exited through the walls in the appropriate directions. Since many of these control wires ran into the reactor building directly adjoining the spreader room, it was crucial that the holes around the cables be sealed airtight.

At 12:20 P.M. on Saturday, March 22, 1975, a work crew was sealing the holes after running several new cables through some of the holes, called "cable windows," in the walls of the spreader room. The standard procedure was to stuff small cubes of polyurethane foam into the holes. Remaining leaks were then sealed by spraying liquid polyurethane foam on the hole. If this combination proved to be airtight, it was then sealed with a flame retardant chemical.

Before applying the last coat, the work crews first had to verify that the foam seal was airtight. To do this they used an old and long-established electrician's trick: they held a candle to the foam to see if the flame was sucked toward the hole by the lower atmospheric pressure in the reactor building on the other side. At 12:20 P.M., an engineer's aide and an electrician were testing the seals around new cables installed in a massive cable window. The aide climbed up on the cable trays and held a candle about one inch from a new seal that had a significant leak.[1] The flame was sucked into the hole and the foam ignited, as it is extremely prone to do. The aide first tried to extinguish the smoldering foam by beating it with the electrician's flashlight. A nearby worker brought some rags to smother the fire, but the rags were soon also smoldering.

Within one and one-half minutes, a fire extinguisher arrived and was emptied on the fire. The fire appeared to be out, but restarted within a minute. Since the fire was smoldering deep within a cable tray, it was difficult to suffocate. As a result, the fire appeared to be extinguished several times during the day, only to restart within a couple of minutes. The aide also now noticed that the fire had spread through the hole into the reactor building, and two workers left the cable spreader room to fight the fire on the reactor building side.

What followed was the chain of amazing errors that often characterizes near calamities. The next two fire extinguishers to be deployed each gave one puff but were otherwise empty. There was difficulty reporting the fire because the posted telephone number was incorrect. There was an automatic CO_2 extinguisher system built into the room. However, there was a delay in using it since a metal plate had been welded behind the breakable glass to keep the workers on Unit III wiring jobs from accidentally suffocating themselves.

The situation on the reactor side of the fire was far more serious. The cable tray entered the reactor building twenty feet above the floor in a room where ventilation was poor. One worker climbed a ladder with an extinguisher but was overcome by smoke. Airpacks to provide oxygen while fighting the fire took five minutes to arrive and then were good only for short periods.

Over the next several hours, the entire five floors of the reactor building had to be evacuated. All lighting was lost in the building at 1:30 P.M. A worker disappeared and was presumed stuck in the elevator. A lengthy search, diverting significant numbers of personnel from fighting the fire, finally revealed that he was standing outside watching the show. Fire fighting had to be stopped for lack of visibility until 2:30 P.M. Then the smoke was cleared, and a five foot visibility was restored by the risky maneuver of opening the door to the second reactor building. The fire was still fought without lights, using a guide rope until a small string of lights was found and strung across the floor.

By early evening the engineers decided they had to fight the fire with water and risk shorting out more electrical cables. The cables would eventually short out from the fire anyway. But having decided to try this, they discovered that the fire department nozzle threads did not match the available hoses. Without the nozzles, the water fell short.

For seven hours, the fire spread despite the best efforts of fire fighters to extinguish it. During the fire, the cable insulations burned and shorted 1600 cables. As the cables shorted, the attached components of the two boiling water reactors were usually rendered useless.

During the crisis, the operators lost control of the pressure relief valves, the main feedwater pumps, the high pressure injection system, the low pressure makeup pumps, the residual heat removal system, and most of the radiation monitors. For a substantial period, water was pumped to the core by the control rod drive motors, which were cleverly rerouted so that they could pump core water. Several people manually operated controls near the core vessel realizing that no one really knew that the core was covered and that the radiation monitors were out of commission.

During the crisis, the Brown's Ferry crew received extraordinary cooperation from government agencies. The Athens, Alabama, volunteer fire department was there almost immediately.[2] After prodding the company engineers all day, they finally put the fire out at 7:20

P.M. when they convinced the engineers that they could rig a water system that would put out even an electrical fire.

The state government in Montgomery was on alert and began collecting air samples within an hour and a half of notification.[3] The Alabama Civil Defense was also on alert by this time, although it was never deployed.

Brown's Ferry is operated by the Tennessee Valley Authority, and the TVA Central Emergency Control Center monitored and advised as the crisis proceeded. TVA was notified at 2:40 P.M. By 3:25 P.M. they had activated the Central Emergency Control Center in Chattanooga.[4] Over the next several hours, TVA maintained communications with authorities in the states of Alabama and Tennessee, the NRC, and the plant. They even eventually gave the authorization to use water on the fire.[5]

Communication with the NRC, however, was time consuming and often frustrating for most parties and provided no obvious benefit to anyone fighting or monitoring the crisis. The NRC maintained offices for Region II in Atlanta. At 3:20 P.M. the utility called the regional office. However, they reached only the answering service. Whenever possible, the federal government does not work on Saturdays.

For almost an hour, the answering service attempted to pass the message to the local office of inspection and enforcement, which would be most directly affected. That notification was finally made some time after 4:00 P.M.[6] Once notified, the responsibility for assembling a team and sending it to the site rested with Frank Long. However, he did not decide that a team was appropriate until 6:29 P.M.[7] The team of three finally arrived at Brown's Ferry about midnight, long after the fire was out and most of the emergency controls had been restored.

In the intervening hours, the NRC abdicated the role of collector or coordinator of information to the TVA. The NRC was almost impossible to reach. The TVA Central Emergency Control Center had individual phone numbers for the NRC officials, and J.R. Calhoun reached Frank Long at NRC at 4:45 P.M., 9:00 P.M., and 10:15 P.M.[8]

Those going through the answering service for NRC, however, were out of luck. These included the Assistant Director of Environmental Planning at TVA, who called at 5:03 P.M. The answering service said he would get a return call within ten minutes. At 5:25 P.M. he called back and was told the service would contact Charles Murphy at NRC at 6:30 P.M.[9] He is still waiting for the return call.

Aubrey Godwin at the Alabama Division of Radiation Health called the NRC at 5:15 P.M. and also reached the answering service.[10] Not receiving a return call, he tried again at 9:30 P.M. Some time after 10:00 P.M., NRC Regional Director Norman Moseley called back to relay information Godwin had already received from the TVA, which was also Moseley's source.

Of course, it can be argued in NRC's favor that they were not in the business of coordinating communications during an emergency and that their proper role was to arrive after the crisis to assess what could be learned to prevent future occurrences. To counteract that argument, much of what the NRC changed after Brown's Ferry was designed to improve its role in the coordination of information. However, even if that was not considered to be NRC's role, the NRC was certainly in charge of monitoring plant procedures over time to protect the public safety. On this responsibility also, NRC's performance relating to Brown's Ferry is open to question.

Polyurethane foam used as a sealant has the significant advantage that it plugs air leaks while being easy to manipulate. But it has the disadvantage that it is extremely susceptible to fire. As early as February 1, 1965, there was a serious cable fire at the then under construction Peach Bottom plant. In that case, the foam was ignited by a welding spark. But most of the fire difficulties with polyurethane foam have been related to the extremely primitive but widely used candle test for leaks. San Onofre Unit I had two such fires in 1968. The Salem, New Jersey, Unit I had a similar fire in April 1974.

But the most embarrassing incident for the NRC happened at Brown's Ferry. On March 20, 1975, just two days before the major fire, a worker ignited some of the foam in the very same cable spreader room with a candle. In that case, he was able to snuff out the smoldering fire with his gloved hand. As required, an incident report describing the event was submitted to the NRC.

At any time, the NRC could have issued a simple order banning the use of candles to check for air leaks in polyurethane foam. Alternative methods were available, and there were numerous reported incidents to indicate that the practice was unsafe. However, the order was not issued until after the Brown's Ferry incident. In a later chapter, there is some speculation on how such a delay might have occurred.

The NRC did some soul searching after Brown's Ferry and completely revised its emergency response mechanism. They did not have the budget or the expertise to create crack response teams ready to fly to the scene of a crisis in progress. Reactor control rooms vary too much for such generalists to be more effective in a crisis than the operators who live with the particular control room daily. However, there was one practice that evolved out of the Brown's Ferry incident that seemed to the NRC to be worth saving.

The NRC has offices in Washington, D.C., although most of the regulatory work is done in the NRC offices in the Washington suburb of Bethesda, Maryland. As the day progressed on March 22, 1975, and the number of inoperative emergency systems at Brown's Ferry grew, personnel from the divisions of Inspection and Enforcement, Nuclear Reactor Regulation, and Public Affairs gathered around the communications room in Bethesda.[11] Although this gathering contributed no decisions or significant assistance during the Brown's Ferry incident, it did serve to keep the high level administrators informed on the progress of the crisis response. It was also the beginning of what by Three Mile Island was called the incident response team.

By 1979, the disaster management apparatus looked impressive compared to 1975. In fact, it was even activated once in the intervening years. But the big test of the new arrangements, on the order of a Brown's Ferry, came with Three Mile Island.

By the new procedures, calls from an endangered reactor to the NRC go first to the regional headquarters. Pennsylvania is in Region I, with headquarters in King of Prussia, Pennsylvania. The first call from TMI was logged in at King of Prussia at 7:04 A.M. The office is not open at that hour, and the call was routed to the answering service.

However, the answering service is much better equipped to handle emergencies than it was in 1975. Even though no one may be in the office, each region is required to have a "duty officer" on call at all times, and the duty officer has the list of phone numbers necessary to swing a response team into action. The answering service operator called the duty officer at home immediately after receiving the call.[12] However, the duty officer was busy preparing to go to work and did not hear the phone. The operator paged him using his "beeper," but he did not hear that either.

At 7:20 A.M. the operator called the duty officer at home again and talked to his wife, who said he was on the way to work. The operator paged him again. This time he heard the beeper. However, since he was in his car on the way to work, he decided to call for the message from there. At 7:44 A.M. Met Ed called the answering service to update the report to a general emergency.

The office was due to open at 8 A.M. At 7:40 A.M. the switchboard operator arrived early and within minutes called the answering service for messages. On hearing of the site and now general emergency, the operator began making the appropriate calls.

Within NRC, the direct regulation of nuclear powerplants is done by the Inspection and Enforcement Office, commonly abbreviated IE. There is a subunit of this office in each regional headquarters, and the regional IE Office maintains an incident response team staffed by people who normally have other responsibilities. The IE team is supposed to monitor events until the national apparatus is in operation, and this was the team first assembled for Three Mile Island.

The responsibility for organizing this team was assigned by Regional Director Boyce Grier to staff member Charles Gallina when he arrived about 8:10 A.M. Gallina assured that open telephone lines were established to TMI and to the national management center in Bethesda, Maryland. By 8:45 A.M., he had gathered a site team including himself, Reactor Inspector James Higgins, and Health Inspectors Donald Neely, Ronald Nimitz, and Carl Plumlee. They drove to TMI and arrived at the site about 10:00 A.M.

The local IE team is supposed to be the first NRC team on site, and it was. However, the meat of the disaster management process that was set up after Brown's Ferry is in the NRC's national offices in Bethesda, Maryland. As the Region I personnel were leaving for Three Mile Island, the incident response program in Bethesda was beginning operation.

The Bethesda office also opened at 8:00 A.M., at which time the public affairs office notified IE Acting Director John Davis that there was a site emergency at TMI. Davis asked Norman Moseley, who had been at Region II during Brown's Ferry and was now Director of Reactor Operations for IE, to verify the story and get more information.[13] Moseley called Boyce Grier, Director of Region I. Grier could verify the site emergency (he had not yet been informed of the general emergency), but he knew no other details.

Moseley then began to assemble the emergency management team, the national equivalent to the team Gallina was simultaneously assembling at Region I. By 8:50 A.M., Moseley had gathered himself, Davis, Vic Stello from the Division of Operating Reactors, and several others at their incident response center. Under the title of the supportive Incident Response Action Coordinating Team, Moseley had also gathered representatives from each of the appropriate NRC offices. Several other people, notably Stello's assistant, Darrell Eisenhut, stayed in their offices to coordinate the activities of these special teams of experts. Eisenhut in particular gathered experts on systems performance and radiological issues in his office to be on call to the coordinating team.[14]

As these people gathered in the center and the surrounding offices, they were in contact with the TMI-II control room and the State Bureau of Radiation Protection through telephone connections to Region I headquarters. At first, beginning about 8:00 A.M., the phone route was open lines from Bethesda to Region I, from Region I to the TMI-I control room, and then from TMI-I to the TMI-II control room.[15] This arrangement was clearly awkward. However, within an hour, open lines to Region I were added to Unit II.[16] As a result, by 9:00 A.M., the NRC coordinating team began to receive information and attempted to analyze the condition of the reactor.

Back at the plant, Miller now had the crew well organized and working at their assigned tasks. Zewe was most immediately in charge of monitoring the controls. Miller decided that Michael Ross, who was normally the operations supervisor for Unit I, was the best qualified operator available, and he was put in charge of supervising the control room.[17] This freed Miller to pursue whatever activities seemed appropriate.

The appropriate activities until about 8:30 A.M. were to finish implementing the emergency plan and to monitor the reactor. The plan involved a long list of notifications, several of whom had to be called two or three times.[18] It also involved dispatching radiation monitor teams and requesting support services from appropriate companies and local facilities. By 8:30 A.M., this activity was subsiding.

In the reactor, Miller realized by 7:10 A.M. that there were steam voids throughout the primary system. There were several indications of this. Water could not be forced through the hot legs. When an attempt was made to start the reactor coolant pumps, they energized to about one-sixth their normal level, indicating that they were pumping steam.[19]

This introduced the possibility that the core might be partially uncovered. But there were no water level indicators in the core. Any penetrations into the core vessel for indicators are expensive. When the plant was being designed, it made no sense to install an indicator for a level that should always remain full.

To determine by other means whether the core was still covered, Miller became somewhat creative. An uncovered core would clearly be superheated and would register that way on the fifty-two thermocouples placed above the core. But those thermocouple readings had been disregarded for hours, since most of them had topped their scales at 720°F. These scales were not designed to be useful in emergency conditions, and it was not clear whether they were still operational.

But Miller knew that the thermocouples themselves were usable to temperatures in excess of 2000°F. Remembering a trick he had learned earlier in his career, he asked Instrumentation and Control Engineer Ivan Porter if there was any other place where the thermocouples could be read. Porter knew that their cables entered the back of the computer panel downstairs in the cable spreader room.[20]

Porter and four other technicians went to the cable spreader room with a "fluke thermocouple reader" and began testing the readings across the incoming cables. Two were unacceptably low considering the hot leg temperatures, registering about 200°F. But the readings that caused the later controversy were two in the neighborhood of 2400°F.[21] When the information was relayed to Miller by Porter, Miller understood that there was a reading of zero, two readings around 200°F, and one of 2400°F.[22]

Shortly before Porter left the spreader room, technician Skip Bennett suggested to Porter that the readings could be taken faster with a digital voltmeter, using a chart to convert voltages to temperatures. It was not as accurate a measuring technique, but Porter did not care, since he did not believe the readings anyway.[23] Shortly after the new readings began, Porter left to relay the first readings to Miller.

Among the four technicians working with Porter, there was one who stated after they recorded the first readings that he thought the core was uncovered. He has been promised confidentiality by the NRC in exchange for his testimony, but the part of his story recounted here is substantiated by the other people in the room.[24] At first, the other technicians assumed along with Porter that all the readings were unreliable.

But as the voltage readings were taken across all the thermocouples in the next half hour, they fell into the same pattern of extremely low or extremely high readings. By the time the readings were completed, the unnamed technician had the others convinced that the low readings were perhaps malfunctioning because of earlier excess heat, but that the high readings had to be accurate. Technician Skip Bennett went to tell Porter.

Bennett remembers that he found Porter and told him their conclusion. Porter does not remember whether the conversation ever occurred. Any chance Bennett had to convince Porter or Miller that the core was uncovered disappeared within a couple of minutes when radiation alarms sounded. The technicians had to evacuate the area, leaving their completed chart of thermocouple readings in the spreader room.

We now know that the top few feet of the core were uncovered at this point. We will never know how hot the fuel became, although it had to reach perhaps 4000°F to deteriorate as quickly as it did.[25] Guesses of 6000°F have also been discussed by some technicians. Clearly, many of the thermocouple connections in the core malfunctioned, giving Miller his inconsistent readings.

Miller decided that his readings were unreliable, and he dismissed them. The NRC was not told of the 2400°F reading until the event was reconstructed a couple of days later. With readings so diverse, Miller saw no reason to believe any of them.

Having no reliable way to tell if the core was covered, Miller was left with no choice except to continue to push water through the core with the tools he had left. He personally believed the core was covered, although that incorrect assumption did not cause him to do anything at first to complicate the recovery process.

With the reactor coolant pumps off, water was being added through the makeup pumps, using water from the borated water storage tank located outside the main buildings of the plant. Unlike the coolant in the rest of the primary loop, this was still water and not steam, and the pumps could move it. There were a few hundred thousand gallons still available, although it was unsettling to have to use a source that was eventually exhaustible.[26]

There was also the question of what to do with the borated water. It would not circulate through the steam voids. It could theoretically be let down to tanks in the auxiliary building. However, the water was highly radioactive, and there was limited sealed storage space. Radioactive water was already causing releases in the auxiliary building. Therefore, the borated water was being let down through a series of valves and tanks that eventually ended on the floor of the containment building.

Over the next several hours, Miller tried to use the makeup pumps to drive the steam voids out of the system. He knew that he had enough water and dumping space to take several hours if necessary. However, it was frustrating that each time he checked the reactor coolant pumps to see if he was making progress, the pumps were still filled with steam. Between 9:00 and 11:00 A.M., he drove the primary pressure from 1250 to 2100 psi.[27] Voids at that pressure seemed absurd. But still the pumps could draw no water.

The reason for the voids that escaped everyone's attention was that the core was partially uncovered and would stay partially uncovered for most of the rest of the day.[28] Pressures of 2100 psi were acceptable for normal operating temperatures, but the super-heated core elevated the coolant temperatures. At these temperatures, they would have to double the pressures to stabilize the plant, and that was impossible in this plant. Therefore, without knowing why the steam was still there, Miller continued to fight it. As he did, information was being passed to Bethesda through the phone connections. However, they were just beginning to collect information in Bethesda and were not yet in any position to make informed analyses or recommendations.

There was one other early source of technical information. As mentioned earlier, Supervisor of Operations James Floyd was at Lynchburg, Virginia, when the accident started, taking a refresher course at the Babcock and Wilcox simulator. He was Miller's first choice to handle the plant during this crisis, but he obviously was not available.[29] Nevertheless, inaccurate rumors about the accident reached Floyd while he was having breakfast that morning.

About 7:30 A.M., Floyd reached the control room by phone. He was told that the emergency feedwater had been delayed ten minutes, that many radiation monitors were off-scale, and that the radiation dome monitor was recording 80,000–90,000 rems per hour.[30] This last "reading" was delivered correctly, although it is now commonly accepted that the lead shield on the monitor was ruptured and that the reading was one hundred times too high. Also, Floyd was inadvertently not told at first about the open pressure relief valve.

Floyd and a growing crowd of Babcock and Wilcox engineers tried to duplicate the accident on the simulator so that they could relay advice back to the plant. They could not approach a condition being described by the operators. They called the plant again at 9:30 A.M. and were told that the pressure relief valve had stuck open. They were not told how long it had been open since the operator relaying the message did not know.

The people in Lynchburg tried simulating the event holding the valve open for what seemed to be an unreasonably long time of fifteen to twenty minutes. However, they were unable to get the emergency cooling system to engage. Also, when they reintroduced water, the coolant loop programmed into the simulator immediately stabilized.

Floyd and the others never did simulate the conditions that actually caused the problems at the plant—or at least not on the first day. When the conditions were tried later, it was discovered that the simulator had not been programmed to respond correctly. The industry was not prepared to understand the accident in progress. It should also be noted that the external transmittal of information from the plant, even to a fellow employee who was offering help, was operating poorly.

One other activity directly related to the plant was in progress before the NRC arrived. Richard Dubiel, who was Director of Chemistry and Health Physics at TMI, had been assigned by Miller to monitor radioactive releases once the emergency was declared. Dubiel staffed his teams by retaining the current personnel and adding the people arriving for the shift starting at 7:00 A.M. By 8:00 A.M., personnel with meters were stationed on the west shore of the island and at the nearby Yorkhaven Power Station. Another mobile team in a car was traveling up and down Route 441, which runs adjacent to the river. A helicopter had been requested from the state police to monitor at Goldsboro across the river, but it had not yet arrived.[31]

When radiation is released from a site, most of it is usually in the form of radioactive particulate matter, which is blown by the wind. It forms a "plume" or funnel, which slowly expands in the direction that the wind is blowing. However, the wind at Three Mile Island this morning was not cooperating. It was officially designated as one mile per hour in a north/east/south/west direction. Rather than dissipating in a predictable direction, puffs of radiation when released floated aimlessly in high concentrations around the island.

This caused alarm more than once. The highest reading picked up by the mobile team on the turnpike was 10-13 millirems per hour. This reading was considerably later in the crisis and was a cause for notice but not alarm. This rate is equivalent to a chest x-ray after about two to three hours exposure.

But a puff of 50-70 millirems was reported quite early in the accident at the north entrance guard station. Ten minutes later, when more sophisticated monitoring equipment arrived, the puff had moved off in an unknown direction. The guard station now monitored no radiation at all.[32] Later in the day, this sequence was restaged at several places throughout the plant. However, Dubiel understood the nature of the problem and soon was responding as needed without any sense of crisis.

By 10:00 A.M., radioactive releases from the plant were moderate. Monitoring teams were properly stationed and more help was on the way. In the control room, Miller had in excess of twenty people analyzing the situation in the reactor.[33] Because of this crowd in the relatively cramped control room, Ross was busy keeping paths open for the operators working the controls.

At 10:00 A.M., the primary coolant pressure was increasing because of the operating makeup pumps. The reactor coolant pumps were off because of steam in the pumps. It was clear that some fuel damage had occurred, melting some thermocouple probes. Miller believed that the core was now hot but covered. The utility to this point had managed the crisis entirely on its own and had relayed the information just summarized to the still confused group in Bethesda. The core was in fact partially uncovered, although that was not yet understood at Three Mile Island or in Bethesda.

To this point, six hours after the transient and almost three hours after notifications began, the NRC had attempted no input into decision-making on handling the crisis. But this was not to be another Brown's Ferry. This time the NRC intended to help as much as possible. A car appeared at the north bridge to request entrance to the island. The team of experts from NRC Region I had arrived.

NOTES

1. Unless otherwise noted, details from the Brown's Ferry fire are taken from "Final Report of Preliminary Investigating Committee," May 7, 1975, reprinted in *Brown's Ferry Nuclear Plant Fire*, hearings before the Joint Committee on Atomic Energy, U.S. Congress, 94th Cong., 1st sess., September 16, 1975, pt. 1, pp. 144–52.

2. The fire department was called at 1:00 P.M. and was deployed in the plant by 1:30 P.M. Testimony of Donald Knuth of NRC, *Brown's Ferry Nuclear Plant Fire*, p. 21.

3. Aubrey Godwin, Director of the Division of Radiation Health in Alabama, was notified by the utility at 3:20 P.M. After alerting several state officials and asking the EPA to stand by, he had monitoring teams in the field at 5:00 P.M. Testimony of Aubrey Godwin, *Brown's Ferry Nuclear Plant Fire*, pp. 161–62.

4. Tennessee Valley Authority, "Fire at Brown's Ferry Nuclear Plant: Tennessee Valley Authority: March 22, 1975," Final Report, May 7, 1975, pp. 52–53.

5. Testimony of J.R. Calhoun of TVA, *Brown's Ferry Nuclear Plant Fire*, p. 441.

6. Testimony of Donald Knuth, *Brown's Ferry Nuclear Plant Fire*, p. 22.

7. Testimony of J.R. Calhoun, *Brown's Ferry Nuclear Plant Fire*, p. 441.

8. Ibid., pp. 441–42.

9. Testimony of unidentified Assistant Director of Environmental Planning at TVA, *Brown's Ferry Nuclear Plant Fire*, p. 443.

10. Testimony of Aubrey Godwin, *Brown's Ferry Nuclear Plant Fire*, p. 163.

11. Testimony of Donald Knuth, *Brown's Ferry Nuclear Plant Fire*, p. 22.

12. Duty officer scenario from testimony of Boyce Grier, Director of NRC Region I, before President's Commission on the Accident at Three Mile Island, May 31, 1979, pp. 240–41; and "Report of the Office of Chief Counsel on Emergency Response" to President's Commission, October 1979, pp. 6ff. Note that there is a different scenario blaming an inoperative beeper in "Investigation

into the March 28, 1979 Three Mile Island Accident by Office of Inspection and Enforcement," NUREG 0600 (Washington, D.C.: Nuclear Regulatory Commission, August 1979), pp. (IA-49)-(IA-52).

13. Testimony of John Davis, President's Commission, May 31, 1979, p. 243.

14. Testimony of Darrell Eisenhut before U.S. Congress, House, Committee on Interior and Insular Affairs, *Accident at the Three Mile Island Nuclear Powerplant, Part I*, 96th Cong., 1st sess., May 9, 1979, p. 7.

15. Testimony of Gary Miller, President's Commission, May 31, 1979, p. 86; "Report of the Office of Chief Counsel on the Nuclear Regulatory Commission" to the President's Commission, October 1979, p. 205.

16. Testimony of Gary Miller, President's Commission, p. 11.

17. Ibid., p. 122.

18. For a list of calls, see *Accident at the Three Mile Island, Part II*, pp. 263-64.

19. Testimony of Gary Miller, *Accident at the Three Mile Island, Part I*, May 11, 1979, p. 185.

20. Deposition of Ivan Porter, President's Commission. p. 20.

21. Ibid., p. 21.

22. Testimony of Gary Miller, *Accident at the Three Mile Island, Part I*, May 11, 1979, p. 186.

23. Deposition of Ivan Porter, President's Commission, pp. 22-23.

24. "Report of the Office of Chief Counsel on the Role of the Managing Utility and Its Suppliers" to President's Commission, October 1979, pp. 205-09.

25. Testimony of Roger Mattson, *Accident at the Three Mile Island, Part I*, May 9, 1979, p. 15.

26. Testimony of Gary Miller, *Accident at the Three Mile Island, Part I*, May 11, 1979, p. 186.

27. Scenario by NRC reprinted in *Three Mile Island Nuclear Powerplant Accident*, hearings by Subcommittee on Nuclear Regulation of Committee on Environment and Public Works, U.S. Senate, 96th Cong., 1st sess., April 10, 1979, p. 14.

28. Testimony of Roger Mattson, President's Commission, June 1, 1979, p. 72.

29. Testimony of Gary Miller, President's Commission, May 31, 1979, p. 122.

30. For sequence, see testimony of James Floyd, President's Commission, May 31, 1979, pp. 146-54.

31. Testimony of Daniel F. Dunn, commissioner of State Police, Pennsylvania Select Committee on Three Mile Island, Pennsylvania Legislature, August 21, 1979, p. 11.

32. Difficulties recounted by Richard Dubiel, President's Commission, May 31, 1979, pp. 75-77.

33. Testimony of James Higgins, *Accident at the Three Mile Island, Part I*, May 10, 1979, pp. 95-96.

❋   *Chapter 6*

The NRC on Site

As the five NRC employees drove along Route 441 the last few miles to reach the plant, there was no evidence of trouble.[1] The townspeople and farmers were going about their normal business. In the car were five passengers chosen to have expertise in any radiological emergency that might arise. In fact, Charles Gallina had been the inspector in charge during the licensing of Unit I.

They had come prepared for the worst. The car contained radiation-monitoring equipment, respirators, and other emergency apparatus that might be needed. But approaching the plant, the passengers could see nothing wrong.

The first hint of trouble was at the gate. There is a security checkpoint as cars turn off Route 441 and just before the bridge that connects the mainland to the island. They reached that gate at almost exactly 10:00 A.M. and noticed an unusual amount of commotion and confusion among the employees. Several employees had been denied entrance, and the gate was chained shut.[2]

Without significant delay, they were permitted onto the island to go to the processing center. Less than half of the island is actually within the security fence. While cars cannot enter from either bridge, local fishermen often dock and even have parties outside the fenced area. There are even some deer and other wildlife that live in a wooded area on the north end of the island.

But the processing center is the point at which intense security begins. Here employees are identified, searched, and checked for prohibited substances and prior exposure to radiation. The NRC inspectors were processed through without delay and were then escorted to the massive complex of contiguous buildings that house the reactors.

On the way to the buildings, they noticed a second sign of trouble. At 10:00 A.M. on a weekday, the grounds of a reactor are normally alive with activity. But this was shortly after the scare at the Unit I security gate. For all practical purposes, the plant seemed deserted.

They were escorted to the closer Unit I control room first, and they arrived there at about 10:20 A.M. James Seelinger, who in normal times was Unit I Supervisor, had just returned from Unit II to greet them. He quickly reviewed what he knew of the status of the reactor, including the steam voids, pressures, temperatures, radiation readings, and the inoperative reactor coolant pumps.[3] There was not time to review how the accident occurred. Seelinger himself did not know what had happened before he went to Unit II at 8:00 A.M.

Seelinger expressed two concerns: First, normal cooling must be restored to the core. For this, Higgins and Neely would return with Seelinger to Unit II. Second, radiation monitoring needed to be intensified to discover what releases were in progress and where the radioactive clouds had floated. For this Plumlee and Nimitz would lead monitoring teams in the field while Gallina would stay in the plant to coordinate communications. Seelinger asked Gallina to stay in the control center behind the Unit I control room since the crowd in Unit II was getting out of hand. Crowd control remained a problem in the control room of Unit II throughout the day. Several Babcock and Wilcox employees who arrived later were not admitted to the plant at all.[4]

As Higgins and Neely headed for Unit II, there was another ominous sign. High radiation puffs were meandering around the buildings at Unit II. They could not proceed until they put on respirators. However, they each had one, so they were soon on their way.

While Higgins and Neely knew about what to expect, the scene in the Unit II control room was unsettling. About twenty to thirty people were milling around in respirators. These were a mixture of operators, engineers, and maintenance personnel. Talking was difficult in the respirators and talking on a telephone was almost impossible. As a result, casually discussing the history of the transient was out of the question. Higgins gathered what information he could on the current status of the plant and began relaying it back to the incident response center at Region I. From there, information was relayed to Bethesda.

Bethesda's sources of information were clearly strained. They received and transmitted information through NRC personnel on open phone lines at King of Prussia. To discuss reactor problems, King of Prussia had an open line to Higgins at Unit II, but he was in a respirator for the next few hours. To discuss radiation problems, Region I had an open line to Gallina in Unit I. However, as time went on, his information on the status of the plant became more out of date.

Within the reactor, Miller was bringing the pressure up to normal levels to try to collapse the voids. By 11:00 A.M. he had reached normal operating pressures and was making no obvious progress. This presented him with two potential problems. First, while he had plenty of borated water, he wanted to use it to make progress. If the plant stayed in this unstable condition indefinitely, he would eventually run out of water and let-down space, and he would have to find less desirable substitutes for his current sources.

But very much in the front of his mind was another more immediate danger.[5] To regulate pressure with containment isolated, Miller had been periodically opening the block valve beyond the stuck pressure relief valve. In fact, an earlier such release had caused the containment building to automatically isolate just before 8:00 A.M.[6]

This valve was normally the back-up system for the pressure relief valve, but it was now in direct line to the coolant water and Miller knew that it had a history of failures. Should it fail now, there was no back-up. The contents of the primary loop would empty, or "blow down," and cooling the core would become a monumental challenge.

About this time, Raymond Smith and Walt Baunack arrived at the north gate from NRC Region I.[7] However, they would not be able to get to Unit II for quite a while, and that was where the important decision of the moment was being discussed.

Miller wanted to begin a controlled depressurization of the primary system while "controlled" was still an option. By hindsight, the recommendation was dangerous. With steam and hydrogen filling much of the core vessel, depressurization of the water would allow the "void" to push more water out of the way and uncover more of the core. However, Miller believed that the core was still covered and never considered the possibility that hydrogen might be in the core.[8] Besides, as risky as depressurization might be, Miller did not know what else to try.

The idea was discussed by numerous people in the control room through their respirators from 11:00 A.M. to almost 11:30. The goal was to open the pressure relief block valve and blow the pressure down to no more than 400 psi. A different set of pumps, the residual heat pumps, could then be used. They can pump a much larger volume of water than can the makeup pumps. With any luck, the operators hoped that the larger volume of water might break through the steam voids and get a flow through the generators. If the tactic worked, the system could circulate its own water and meet the immediate objective of stability.

Higgins called Region I to get Bethesda's advice on the matter. The center at Bethesda was now very crowded and overflowed into several rooms. To avoid a feeling of chaos, the emergency management team had isolated themselves in one room in the center. The advisors now available at Bethesda were numerous. But given future events, there were two notable absences. Harold Denton was still in his office and remained there until 5:00 P.M.[9] He was not involved in emergency management, but had sent an assistant to watch.

Also, three of the five NRC commissioners were missing. Chairman Hendrie was at the Washington Hospital Center, where his daughter was undergoing surgery.[10] Richard Kennedy stayed in his office that day. Victor Gilinsky called a friend at the White House that morning, as discussed later. However, he did not go to the center.

John Aherne was notified about the crisis and went straight to the center, where he spent the day. Peter Bradford went to the center for the novelty; he had never seen it in operation and wanted to see how it worked.[11] He did not leave.

However, despite the numbers now gathered in the room, no one had the information needed to give useful advice. Depressurization to use the residual heat pumps is a normal part of bringing a reactor to a cold shutdown. It is a desired state of affairs. The fact that the reactor operators were considering this move was good news, because it presumably meant that this next step toward a shutdown was possible. No objections were raised in Bethesda.[12]

At 11:30 A.M., Miller ordered the block valve opened to begin depressurizing the primary loop into the containment atmosphere.[13] He also ordered the high pressure makeup pumps throttled, but left on to keep water flowing around the core. Immediately, and for the next thirty minutes, the pressure began to fall as more water and steam escaped than was replaced. From a high of 2100 psi, the system began falling toward the goal of 400 psi.

About noon, the people in the center in Bethesda finally became totally frustrated with the awkward three party system for getting information from Three Mile Island. They asked the people at Region I to have the plant establish a direct open line from the plant to Bethesda. Preparations for this were started.[14]

But the news at noon was in the reactor. For thirty minutes, the pressure had fallen from 2100 to 500 psi. During this period, the core flood tanks, which are available to literally dump water by gravity into the vessel, had partially discharged.[15] This was interpreted as a good indication that the core must be covered, since the discharged flood water showed that water was still moving in the core vessel.

But the depressurization stalled at 500 psi. Miller needed an additional drop of 100 psi before the residual heat removal pumps could push water. For a second time, he faced an unstable plant that could be pushed no further. He was not willing to shut off the high pressure makeup pumps, because of his fear that shutting off the pumps might uncover the core. However, those pumps were now supplying water as fast as steam was leaving. At least this way he was not using his borated water as quickly as he had been when the plant was highly pressurized. Therefore he stalled for time, hoping the pressure would fall another 100 psi.

It is now suspected that this depressurization period uncovered even more of the core as the pressurized steam and hydrogen pushed the decreased water flow lower into the core vessel. At least, the radiation monitors were climbing to indicate this possibility.[16] At the same time, hydrogen was entering containment through the block valve and would later "explode" or "burn," depending on definitions.

To complicate matters even further, new off-site radioactive releases had begun at 11:00 A.M. because of the damaged vent header on the waste gas line leading from the quench tank to the auxiliary building. Once the block valve was opened, the releases logically increased. However, radiation information was monitored in Unit I, and Miller did not know of the off-site releases at this time. Much to his embarrassment later in the day, the state monitoring agencies did discover the releases and measured them continuously from 11:00 A.M. to 1:30 P.M.

For the next several hours, the parameters of the primary loop stayed about the same, with the thermocouples apparently malfunctioning from what was believed to be earlier overheating and the pressure at about 500 psi. However, Miller was about to get introduced to the world of politics, which interfered with managing the depressurization and made the behavior of the reactor look simple by comparison.

With all the activity in the control room, the people in their respirators were largely isolated from the rest of the world. Communications with Bethesda improved in midafternoon when the respirators were finally removed. However, real improvement did not come until late afternoon, when Unit II was finally tied directly to Bethesda by open telephone lines.

But by now the world, through the press, was watching Three Mile Island. The crowd in Unit II was not aware of it, but a sizable contingent of reporters and film crews was gathering at the observation center, located a few hundred yards from the reactor buildings on the closest part of the mainland. Met Ed Vice President John Herbein had arrived from Philadelphia to manage press relations. For that purpose, and because privacy is difficult in the observation center when it is crawling with people, a small trailer had been wired for telephones on the side lawn of the observation center. From the trailer, Herbein could talk to the press.

Miller and his team of workers were not at first aware of this development. However, the world of public relations was to influence his struggle with the reactor twice in the next few hours. The first incident was a mere aggravation. Generator B had long been isolated because of the belief that it was internally ruptured. Generator A was operational, and Miller had satisfied himself that it was not leaking. However, he had problems maintaining an adequate flow of water in the secondary loop during the late morning. It was nothing permanent or serious, but for a while he needed to break the secondary loop.

He routed generator A's secondary steam to the atmospheric dump valves. These exit through a set of sixteen large pipes near the intersection of the containment building and the turbine building. They look very much like the pipes in a huge pipe organ, and they can generate an intense noise. However, so long as there is no primary-secondary coolant breach, they are harmless. To ensure that the valves were not releasing any radioactive material, Miller placed a man on the roof of the turbine building to monitor the discharge.[17]

Over the next hour, Miller was encouraged by several sources to stop the atmospheric dumping if possible. Specifically, Herbein passed the message that the state feared that the plant was releasing radioactive steam. For a while, the dumping could not be stopped. A boiler was out of commission and was needed to create the vacuum that causes the secondary loop to circulate.

But Herbein was getting more political pressure than Miller realized. At 12:30 P.M. Herbein ordered the steam release stopped.[18] Fortunately, some vacuum had been restored, and the secondary steam was successfully routed back to the condensors for normal cooling through the tall cooling towers.

The other public relations problem affected Miller's activities much more heavily. State monitoring agents had been picking up a new radioactive release from the auxiliary building from about 11:00 A.M. to 1:30 P.M. There had been a series of confusing exchanges outside the plant on this subject that are reported in more detail in the next chapter. In a nutshell, the Lieutenant Governor had reported the releases to the press and was then told that John Herbein was claiming that there had been no off-site releases all morning.

Miller had no part in the exchange, knew of no releases since such information was filtered through Unit I, and did not know that a war of information was taking place. Nevertheless, when the Lieutenant Governor summoned Herbein to a 2:30 P.M. meeting in Harrisburg to straighten out what was happening at the plant, Herbein decided to take George Kunder and Gary Miller along. Miller protested.[19] He had nothing to do with public relations, he did not understand the subtleties of politics, and he was more than a little busy at the moment. However, he was "strongly urged" by management to attend, and he finally did.

Shortly before he left, the entire control room was given an unmistakable sign that the damage to the core was far more serious than anyone imagined. Miller sincerely believed before this point that he was 100 psi from a stable plant and the end of the crisis. We now guess that the core was uncovered to a distance of perhaps four feet. But with the pressure decreasing and the thermocouples assumed out of commission, the readings did not indicate a crisis.

But at 1:50 P.M., the indication came. During the past several hours, the heat had been baking the cooling water into its hydrogen and oxygen components. The oxygen had been combining with the zirconium, increasing the radioactivity as the zirconium coatings on the core deteriorated. The hydrogen had been collecting in the void at the top of the core vessel. Once the block valve was opened at 11:30 A.M., it had begun escaping through the pressurizer into the containment atmosphere.

At 1:50 P.M., the hydrogen in the containment building finally recombined in a process that was halfway between an explosion and a slow burn. It lasted several minutes and drove the containment building pressure up to 28 psi.[20] The building is only rated to withstand a 60 psi internal pressurization, although it can withstand much higher or lower pressures depending on whether there is a shock wave of explosive force.[21] The containment building was in no danger. Nevertheless, the burn indicated that a very large amount of hydrogen had been created, which could happen only if the core was substantially uncovered.

Recollections of the event vary from person to person in the control room. There was a low, muffled boom in the control room. Edward Frederick happened to be watching the containment building pressure strip chart and called Craig Faust over to see the pressure "spike."[22] But within seconds the pressure began to fall.

Gary Miller heard a thud, but it sounded like a malfunction that he had heard before in the ventilation system for the control room.[23] It did not occur to him that an explosion might have occurred. Both NRC people in the room later testified that they were busy transmitting information to Region I in the shift supervisor's office. With their respirators and phones, they did not even hear the thud.[24]

When the pressure in containment soars, there is a fire-fighting mechanism that automatically engages. A spray ring in the dome dumps a mixture of sodium hydroxide, caustic acids, and other chemicals designed to fight fires and cause most radioactive isotopes to settle out of the air. It is effective if one assumes that things are so unstable in containment that explosions are occurring. The chemicals involved can limit the spread of radioactive isotopes in a containment rupture, for instance. However, the chemicals are extremely caustic and will eventually cause much of the instrumentation in containment to malfunction. Therefore, the system is almost never used and is not even turned on for testing when installed.

Faust, Frederick, and Operations Supervisor Michael Ross watched the system engage. If there was a fire in containment, they wanted the system left on. But the pressure spike began declining before the spray ring was fully discharging. The three discussed the situation briefly, realizing that the spray was unnecessarily damaging equipment if it was on accidentally. They finally decided that the spike had been so short that it must be equipment failure.[25] Therefore, within two minutes of the time it engaged, the spray ring was turned off.

There is some confusion in later testimony on who knew of the incident. Ross told James Floyd two days later that the NRC people were aware of the decision to shut the spray off.[26] Higgins of NRC acknowledges that numerous power fluctuations during the day accidentally started several safety systems that had to be manually stopped.[27] However, he is sure that this was not one of them. Under the strain of the confusion and the respirators, somebody remembered the event incorrectly.

At any rate, the best indication available on the first day that the reactor was in serious trouble was completely missed. Most of the room did not see it, and those who saw it did not believe it. It appeared to be one more in a long line of equipment failures and was not reported to anyone else.

Instead, as Miller prepared to leave for his 2:30 P.M. meeting with Herbein, Kunder, and the Lieutenant Governor, there was some encouraging news. Pressure had decreased to 440 psi, just 40 psi more than he needed to start the residual heat pumps.[28] The core flood tanks had settled about a foot and a half in level, indicating that water was entering the core. And just before Miller left, an experiment that was tried with the flow of the high pressure makeup pump C caused both the hot leg temperatures and the cold leg temperatures to move.

For the first time in hours, Miller had some indication that water had flowed through the primary loop. It was minor, but it was a start. There was the potential now to begin to reach stability. But Miller's time had run out; he had to go to the meeting.

For some time after Miller left, the system began to show increasing signs of improvement. The pressure would fall no more; it bottomed out at 440 psi and began slowly to rise. However, the depressurization had been tried only as a tactic to get water flowing across the core in a continuous loop. But now there was increasing evidence that that flow through the generators was beginning, using just the B makeup pump and fluctuating the flow from the C pump. Since the residual heat pumps were intended to force flow through the generators, if the flow started without them, there was no problem.

Apparently, the voids began to shift with the changing water flow. At 3:06 P.M., the pressurizer level began to fall.[29] It had been full, or within inches of full, since shortly after 8:00 A.M. The level now fell dramatically, despite the open block valve, as the water presumably fell back into the primary loop. Eighteen minutes later it bottomed at 180 inches, which was actually a little below normal operating range. Of course, normal operating figures meant nothing now. But wherever the water was going, the hole replugged at 3:24 P.M., and the pressurizer climbed rapidly back off-scale.

During this drop, some of the water was clearly filling generator A. The dry cold leg rose from 200 to 400°F as it presumably received heated water. With the heat sink in the generator, the hot leg temperature also dropped onto scale at 560°F.

At 4:00 P.M., Miller returned to the plant. The ordeal of that meeting is recalled in a later chapter. But at any rate, he needed some good news now, and the reactor was giving him some. There was no real progress at the moment, but there did appear to be some flow through the system. While the plant was not yet stable, it was a giant step closer.

Also, to improve everyone's spirits, the respirators were now off. While releases were still floating around the plant, they were not at

Unit II. In fact, something of an irony developed shortly after Miller's return, when radiation alarms went off in the Unit I control room. Between 4:00 and 5:00 P.M. the NRC people who had been left in Unit I to minimize the crowd had to be moved to the now safe Unit II control room.[30] Charles Gallina was unhappy about the switch, since the field teams were set up to call Unit I. However, the radiation prevailed.

For most of the day the NRC had two inspectors in Unit II. Now they had several. The team in Bethesda was very large and grew sizably when Darrell Eisenhut moved his technical staff to the incident center about 4:00 P.M.[31] Throughout the day, the NRC had the power to pull out an old section of the Atomic Energy Act and do anything from giving orders to actually seizing the plant in the name of public safety.[32]

However, that was clearly not appropriate. The center in Bethesda asked many questions during the day, which the utility was required to answer. The answers came willingly, and only the communications were bad. But no one in Bethesda entertained the notion that they could operate the plant over the phone.

On the site, the NRC staff had access to any information they wanted. Their biggest hindrances had been the respirators and the necessity to continually call back to headquarters. Squeezed within these constraints, they had become part of the team.[33] When system changes were contemplated, they were bantered around the room to get all the input possible. Higgins and several others were part of the group kicking around the ideas.

The system had worked pretty well. So far, the pressure spike was the only incident of substance missed by the NRC. Given the reaction of those who decided to turn off the spray, it is not clear that Higgins would have handled it any differently. In fact, there is some testimony that Higgins heard of the event at the time.

At any rate, there was now another decision to banter around the room. The depressurization tactic had clearly failed. By 5:00 P.M., the pressure had risen from a low of 440 psi to over 600 psi. But the residual heat pumps might not be needed after all. Using high pressure pumps, a small flow through the steam generators had been started.

Besides, Babcock and Wilcox had called after calculating with James Floyd that the high pressure pumps must maintain a 400 gallon per minute flow to keep the core covered.[34] There had been some concern, although only suspicion, for the past couple of hours that if the thermocouple readings were correct, the core might be uncovered. Gallina from the NRC had been present when Vic Stello had discussed this on the phone.[35]

With the high pressure injection increased to 400 gallons per minute, there was no chance to depressurize. Therefore, the Met Ed personnel, the NRC, and Met Ed management all agreed that the operators should attempt once again to collapse the voids with pressure. Shortly after 5:00 P.M., the block valve on the pressure relief valve was shut, and the pressure began to climb.

Two systems needed to be prepared should the repressurization work. First, normal secondary feedwater flow through steam generator A needed to replace the auxiliary feedwater that had been cooling the generator since eight minutes into the accident. The polisher bypass valve now worked, and the main feedwater pumps were restored to service at 5:08 P.M. The turbines, the polishers, and generator B were bypassed, but the secondary loop was now cooling by circulating its own water in a closed loop through generator A.

The more serious difficulty was that the 10,000 horsepower reactor coolant pumps would not operate unless oil pumps kept them lubricated. With all of the instrumentation that had been knocked off line that day, the oil pumps were out of service. They had to be started, or the reactor coolant pumps could not be used to restore stable cooling.

The pumps themselves were not harmed. But their power supply had shorted, and the circuits were in the radioactive auxiliary building.[36] Fixing them involved a considerable balancing of priorities between what needed to be done to the circuits and how long employees could be exposed to the work areas. The work was done, but it proceeded slowly, and no worker received more than a fraction above his allowable dose.

NRC representation was growing rapidly. By early evening, eleven NRC employees were at the site.[37] The late arrivers came with a mobile radiological laboratory, which had been in Connecticut when the accident started and which arrived at the plant around 6:30 P.M. With it, more sophisticated radiological tests could be performed away from the plant site.

Within the plant, work on the oil pumps progressed while the system repressurized. By 7:30 P.M., an oil pump was finally started, and the pressure was somewhat above operating levels at 2300 psi. The plant was ready to try for stability by starting up a reactor coolant pump.

In the second incident of unusual timing that day, NRC representatives Higgins and Gallina were now called away for a meeting with the Lieutenant Governor.[38] They had plenty of relief available in the control room, and Bethesda was kept informed. But they left within minutes of the planned test, not knowing if the pumps would start.

Miller wanted to try an A pump since the B generator was blocked. He tried pump 2A first, but it would not start.[39] He switched the oil over to pump 1A, energized the pump, and "bumped" it. It started. It only ran for ten seconds before stalling, but both the temperature and pressure readings responded immediately. There was unmistakable coolant flow, if only they could get a pump to start.

When a reactor coolant pump shuts off, it needs fifteen minutes of down time before it can be safely restarted. They did not want to burn their pump out now, so they waited the full fifteen minutes. If it would start now with the coolant flow that they had seen in the ten earlier seconds, the crisis of Three Mile Island might well be over.

At 7:50 P.M. it was time to try again. They pumped the oil, energized the pump, and "bumped" it. It started. This time, it stayed on.

Immediately, the coolant flow began collapsing the steam voids left in the coolant lines. Temperatures throughout the primary loop showed the effects of rapid cooling and soon had dropped and stabilized at 380° F in both loops. Pressure, which had been held artificially high by steam voids blocking the lines, collapsed with those voids. Within minutes, the 2000 psi pressure fell to 1320 psi and stabilized. By all indications the crisis in the reactor at Three Mile Island was now over. While Gallina thought that it might take a day to mop up the spilled radioactive water, the only tasks left now were to bring the reactor to a cold shutdown so repairs could begin.

NOTES

1. Scenario from "Report of the Office of Chief Counsel on Emergency Response" to President's Commission on the Accident at Three Mile Island, October 1979, pp. 14–15.

2. Testimony of James Higgins before U.S. Congress, House, Committee on Interior and Insular Affairs, *Accident at the Three Mile Island Nuclear Powerplant, Part I*, 96th Cong., 1st sess., May 10, 1979, p. 94.

3. Deposition of Dr. Charles Gallina taken by President's Commission, Bethesda, Maryland, August 2, 1979, p. 21.

4. Testimony of Joseph Kelly, President's Commission, July 18, 1979, pp. 29–30.

5. Written submission of Gary Miller, *Accident at the Three Mile Island, Part II*, p. 269.

6. "Technical Staff Analysis Report on Summary Sequence of Events" for President's Commission, October 1979, p. 13.

7. Testimony of Charles Gallina, President's Commision, May 31, 1979, p. 239.

8. Written submission of Gary Miller, *Accident at the Three Mile Island, Part II*, p. 269.

9. Testimony of Harold Denton, President's Commission, May 31, 1979, p. 302.

10. Testimony of Joseph Hendrie, President's Commission, June 1, 1979, p. 118.

11. Testimony of Peter Bradford, President's Commission, June 1, 1979, p. 121.

12. "Report of the Office of Chief Counsel on the Nuclear Regulatory Commission" to President's Commission, October 1979, p. 209.

13. Sequence in *Three Mile Island Nuclear Powerplant Accident*, hearings by Subcommittee on Nuclear Regulation of Committee on Environment and Public Works, U.S. Senate, 96th Cong., 1st sess., April 10, 1979, p. 14.

14. "Report of the Office of Chief Counsel on Emergency Response," p. 16.

15. Scenario in *Three Mile Island Nuclear Powerplant Accident*, p. 14.

16. "Technical Staff Analysis Report on Summary Sequence of Events," p. 13.

17. Written submission of Gary Miller, *Accident at the Three Mile Island, Part II*, p. 269.

18. Ibid., p. 273. "Investigation into the March 28, 1979 Three Mile Island Accident by Office of Inspection and Enforcement," NUREG 0600 (Washington, D.C.: Nuclear Regulatory Commission, August 1979), p. (1A–84), says vacuum was not restored for almost five hours, leaving no secondary heat sink for that period. The other available parameters of the plant do not clear up the contradiction, but neither story affected the difficulty of the transient.

19. "Report of the Office of Chief Counsel on Emergency Response," p. 18.

20. Testimony of Carl Michelson, *Accident at the Three Mile Island, Part I*, May 10, 1979, p. 73.

21. Testimony of Donald Ray of Babcock and Wilcox, President's Commission, July 20, 1979, p. 393.

22. Testimony of Edward Frederick and Craig Faust, *Accident at the Three Mile Island, Part I*, May 11, 1979, p. 143.

23. Testimony of Gary Miller, President's Commission, May 31, 1979, pp. 57 and 60.

24. Letter of May 18, 1979, from John Davis to Lee Gossick, reprinted in *Accident at the Three Mile Island, Part II*, p. 192.

25. Testimony of Michael Ross, President's Commission, May 31, 1979, p. 60.

26. Testimony of James Floyd, President's Commission, May 31, 1979, p. 217.

27. Testimony of James Higgins, *Accident at the Three Mile Island, Part I*, p. 99.

28. Written submission of Gary Miller, *Accident at the Three Mile Island, Part II*, pp. 273–74.

29. Plant parameters from "TMI-2 Interim Operational Sequence of Events as of May 8, 1979," May 8, 1979, pp. 14–15 (mimeographed).

30. Deposition of Charles Gallina to President's Commission, pp. 24–25.

31. Testimony of Darrell Eisenhut, *Accident at the Three Mile Island, Part I*, May 9, 1979, p. 8.

32. Testimony of Norman Moseley, President's Commission, May 31, 1979, p. 251.

33. Testimony of James Higgins, *Accident at the Three Mile Island, Part I*, May 10, 1979, p. 100.

34. Written submission of Gary Miller, *Accident at the Three Mile Island, Part II*, p. 276.

35. Deposition of Charles Gallina to President's Commission, pp. 64–65.

36. Testimony of Gary Miller, *Accident at the Three Mile Island, Part I*, May 11, 1979, p. 187.

37. Testimony of Joseph Hendrie, *Three Mile Island Nuclear Powerplant Accident*, p. 67.

38. Testimony of James Higgins and Charles Gallina, *Accident at the Three Mile Island, Part I*, May 10, 1979, p. 98.

39. Written submission of Gary Miller, *Accident at the Three Mile Island, Part II*, p. 276.

❋　*Chapter 7*

The State Seeks a Role

Back at 6:55 A.M., when the reactor deteriorated into a site emergency, some of the personnel in the control room were ordered into the shift supervisor's office to begin the lengthy process of off-site notification. The NRC, which had no capacity to respond quickly, was called about ten minutes into the notifications. The state was called almost immediately.

To make any judgments about the state's role in the Three Mile Island crisis, or even to recount what they did, it is first useful to understand just how little states are required to do during radiological emergencies. For all practical purposes, states with reactors need not have any plans to respond to accidents.

Neither the NRC nor the older AEC has ever put much faith in state (or federal) emergency plans. Safeguards were required and built into the plants so that minimum attention needed to be drawn to off-site planning. For years, it was the NRC philosophy that "emphasis on radiological emergency planning would serve only to arouse public concern and to stifle the development of nuclear power."[1]

It should not be interpreted from this that the NRC or the AEC was completely flippant on the issue. When Con Ed requested permission to build a reactor in the borough of Queens, for instance, the AEC drew the line. To stop such requests in the future, it set to work in the 1960s writing minimum population requirements for future reactor sites. The requirements AEC created, however, help illustrate the problem.

Prior to the crisis at Three Mile Island, siting restrictions for proposed nuclear reactors were spelled out in 10 CFR 100.3 of the Federal Code. Emergency preparedness requirements were defined in Appendix E of 10 CFR 50. Neither of these gave the state or the utility the detailed information they needed for effective planning. For that, the staff of the NRC issued Regulatory Guide 1.101. The guide was strictly advisory, but the NRC expected that utilities would use it when devising plans to earn an NRC license.

The base unit for all emergency planning around reactors in the United States has been the "low population zone" or LPZ. The LPZ is the maximum distance a person can stand from a nuclear reactor and get a 25 rem whole body dose or a 300 rem thyroid dose from an accident involving multiple safety system failures but no breach of containment. Such a dose might or might not make an individual ill, but is nowhere near the lethal dose for the average human.

The exact distance involved in the LPZ is a factor of the terrain and the safety features in the plant. It varies at each reactor in the United States, from about one-half mile to a little over six miles.[2] At TMI, the LPZ is two miles.

The siting requirement for reactors is that at the time of licensing, no reactor may be placed within one and one-third LPZs of the closest "population center." A "population center" is a town with at least 25,000 people. In other words, TMI had to be almost three miles away from the nearest town with 25,000 people. Had it not been able to meet this requirement, the plant could still have been sited in the same place if its safety features had been improved to decrease the LPZ distance.

Beyond the minimum distance, or after the date of licensing, population may be present in any concentrations. In 1977, about ten million people lived within twenty miles of a reactor.[3] The most extreme examples are at Zion, Illinois, outside Chicago and at Indian Point above New York City.

Indian Point, the site of three reactor cores, is about twenty miles from the Bronx and forty miles from Times Square. Over twenty million people live within fifty miles. Robert Ryan, Director of State Programs for the NRC, guesses that dozens to hundreds of people would have been killed in the panic had the TMI incident happened at Indian Point.[4]

To be constructed, the plant must meet the "population center" minimum distance test at the time of licensing. But the LPZ is also the base unit for the requirements for emergency planning. To get a license, the utility or the state or the two working together must present a plan to demonstrate that they can evacuate the population and

destroy contaminated foodstuffs within the LPZ.[5] Any plans beyond
the LPZ are the sole responsibility of the state authorities or their
designated representatives and are strictly voluntary.

The states are not required by the NRC to have evacuation plans
for reactors to be built, licensed, or operated. Many states before the
Three Mile Island incident did not have plans, including two states
with several reactors each. Pennsylvania, however, did have a plan
that had been written in 1975 and periodically updated since then.

The NRC provides potential assistance for states in devising such a
plan through its Office of State Programs. However, the assistance
function has always been dramatically understaffed. In 1969, when
about twenty reactors were operating in the United States, seven
employees in the AEC devoted a total of about one to two person
years to emergency planning.[6] By the time of Three Mile Island,
among a total of 2500 NRC employees, three professionals and one
secretary worked on emergency planning.[7]

NRC assistance for state emergency planning has been in two
forms. First, beginning in 1974, the NRC suggested the contents of a
good state plan by listing 154 criteria.[8] The criteria were so numerous that no states were able to bring their plans into compliance with
all of them. Of course, there was no requirement that they even try.

Second, while the NRC did not have the authority to "approve"
or "disapprove" state plans, the Office of State Programs did offer
to "concur" in those plans that met the 154 criteria. Unfortunately,
for the first three years, no plans could meet the requirements for
concurrence. Finally, in 1977 the Office of State Programs offered
to concur with plans that met the most important 70 criteria. Almost
immediately, some state plans received concurrence. The first such
state plan was South Carolina's.

It is indicative of the state of information in emergency planning
that there is disagreement on how many states achieved concurrence
before the Three Mile Island accident. The NRC says that it was
eleven.[9] The U.S. General Accounting Office says that it was nine.[10]

There is also confusion on the status of the Pennsylvania plan.
Pennsylvania Emergency Management Director Oran Henderson testified that it had never been submitted for NRC concurrence.[11] Joyce
Freeman, staff assistant in the Lieutenant Governor's office for the
Pennsylvania TMI Commission, remembers that the plan was rejected
once and was being revised for resubmission to the NRC.[12] At any
rate, at the time of the Three Mile Island accident, the Pennsylvania
plan did not have concurrence.

Pennsylvania's entire emergency management structure was revised
in March 1978. That structure was used for all types of emergencies,

including the floods that are common to Pennsylvania and the chemical transport spills that are common to all states. The plan also applied to emergency planning around reactor sites within the state boundaries.

Pennsylvania houses several reactors, with LPZs ranging from two miles to almost five miles. To standardize their plan, the state has prepared evacuation plans for a five mile radius at all of its nuclear sites.

This limited radius decision was challenged during the siting hearings for TMI Unit II. A group called Citizens for a Better Environment appeared without legal counsel.[13] They challenged the assumptions in the five mile plan that all state and local officials were on call around the clock. State and county officials, represented by Met Ed legal counsel, testified that those assumptions were essentially valid. At least for the single case of Three Mile Island, the officials seem to have been largely supported by history.

The citizen's groups also wanted to know what plans existed beyond the five mile radius. Met Ed lawyers successfully stopped the challenge on the grounds that it was not apparent that an evacuation beyond five miles would ever be needed. The challenge was stopped, but the testifying county official, Kevin Molloy, was later to face that five mile decision with far more anxiety than was possible during these hearings. With the major arguments of the opposition counterchallenged or dismissed, the Pennsylvania plan won the only approval really necessary. Three Mile Island Unit II was licensed.

The Pennsylvania emergency plan is rather advanced in that it creates a distinction between, and a rather successful melding of, the usually confused concepts of emergency planning and emergency preparedness. Emergency planning is the process of deciding in advance what decision-making structures and resources will be available when needed and how they are to be "triggered." Emergency planning is begun at the local level, where the best information exists on escape routes, points of traffic congestion, local habits, and so forth. It should not be overlooked that numerous Amish and Mennonite families live within twenty miles of Three Mile Island and that they have no televisions, radios, or telephones. Only a local emergency management director would know how to reach them in an emergency.[14]

Emergency preparedness, leading to emergency response once activated, is the process of maintaining the trigger mechanism and resources in such a state that the plan can be activated when needed.

For this function, there were three offices of importance in Pennsylvania. Colonel Oran Henderson was the Director of the Pennsylvania Emergency Management Agency, commonly called PEMA and equivalent to what is normally called the civil defense office in other states. As a paramilitary leader, Henderson was organizationally equipped to respond to the extraordinary needs for leadership during a crisis.

But more long-term responses to an emergency might well call on the resources of the Governor. The Governor, for instance, must ask for any federal declarations of a state of emergency. He must also issue all evacuation orders except when that is clearly inappropriate during a localized immediate crisis.

The Governor is not directly involved in the emergency action chain of information since there are too many local emergencies and too little gubernatorial time. Instead, the Lieutenant Governor is the head of the State Emergency Council. The Lieutenant Governor serves as the filter of information from PEMA to the Governor on state emergencies. If the emergency requires higher level response, the Governor's office is literally down the hall from the Lieutenant Governor in the state capitol in Harrisburg.

The third office of importance, although this office is often notified first in emergencies, is the emergency management director of the affected county. This office has the advantage that it can be a conduit of information between local emergency management directors who often have the best local information and the state level where there is a broader perspective and more access to emergency resources.

At 6:55 A.M. on March 28, 1979, with the Three Mile Island reactor in trouble, the operators made two calls to the state of Pennsylvania. The first was a call from William Zewe in the control room to Clarence Deller, the duty officer for PEMA. The call, which was recorded in the log at 7:02 A.M., was literally the first call of notification made from the plant.[15]

Zewe told Deller that there had been an emergency resulting in high radiation in the reactor building. Zewe said that there had been no releases off site. To the best of Zewe's knowledge, that was true.

Zewe made no attempt to explain the accident to the PEMA duty officer, since Deller was not trained in nuclear reactor operations. Instead, as the state emergency required, Deller was to call the Bureau of Radiation Protection (BRP) in the Department of Environmental Resources. BRP would call the plant and would have the expertise to determine what state actions were appropriate. BRP would then call PEMA back with a recommendation to pass to the Lieutenant Governor.

A minor snag developed. There was no one at the Bureau of Radiation Protection to answer the phone. However, the PEMA duty officer knew to call BRP's duty officer, William Dornsife, at home.[16] Dornsife, the state government's only full-time nuclear engineer, then began the notification process inside the Bureau of Radiation Protection.

He called Margaret ("Maggie") Reilly, BRP's Chief of the Division of Environmental Radiation. Maggie Reilly is an interesting, personable, and extremely respected member of the nation's radiological community. According to one of the county emergency management directors in Pennsylvania, she almost singlehandedly wrote Pennsylvania's emergency response plan.[17] Whether that is overstated or not, she has a habit of appearing where things are happening. For instance, on March 25, 1975, the Alabama Radiation Health Director received his only call during the entire Brown's Ferry incident that was not from a directly affected party. Maggie Reilly (misspelled in Godwin's notes) wanted to know about the need for and the use of the emergency core cooling system.[18]

Almost exactly four years later, Ms. Reilly was very much directly affected. Dornsife informed her of the site emergency and asked her to establish the required open telephone line to the plant.[19] While Reilly worked on the open line, Dornsife called the TMI switchboard.[20] They could not, or possibly would not, connect Dornsife with the TMI-II control room. However, he left his name and phone number for a return call.

At 7:15 A.M., the control room called Dornsife back. Dornsife was told that there had been a now contained small loss of coolant accident.[21] Within the jargon of the industry, this was an effective quick summation of the problem. He was also told that there was a site emergency because of high radiation in the plant, but that no radiation had been detected outside the building. Finally, he was informed that the plant was now stable and being cooled normally.

The last statement was an interesting interpretation of what the operators knew about the condition of the primary loop at this time. However, Dornsife was not left with any false sense of security. Just before the conversation ended, there was an event that could not have been professionally staged for a better dramatic effect.

In the background of the phone conversation, the TMI loudspeaker blurted out that there was radiation in the fuel-handling auxiliary building and that the building was to be evacuated immediately. At this point, Dornsife said to himself that "this is the biggie."[22] Dornsife's phone connection was immediately transferred to the health physics department, where he was assured that there were no off-site releases.

For several minutes, the state officials stumbled over each other double checking notifications. Margaret Reilly called Dauphin County Emergency Director Kevin Molloy at home to make sure that he had been notified.[23] While they were talking, Molloy's fire monitor signaled him to clear the phone line and call his office. There he received PEMA's message, although the "site emergency" was garbled into a "slight emergency."[24]

Over the next ten minutes, PEMA notified the three affected counties, although there was no answer at first at the York County Emergency Operations Center. PEMA also notified the National Guard, who began cataloguing their available equipment and waiting for further word.[25] The Bureau of Radiation Protection advised PEMA that no evacuation was appropriate. Still, Molloy in Dauphin County began contacting his local directors.

So far, the emergency plan was proceeding with only minor snags. But when the people in the control room began notifying the state to upgrade the site emergency to a general emergency, they made a major error, causing a number of people considerable alarm.

Richard Dubiel, TMI's Director of Chemistry and Health Physics, was receiving astronomical readings in the dome of containment because of what is now commonly accepted to be a ruptured lead shield on the dome's radiation monitor. Based on his calculations, if that reading was reliable, radiation in the town of Goldsboro across the river would be 10 rems per hour. At that level, Goldsboro needed to be evacuated immediately.[26]

Rather than call the Bureau of Radiation Protection directly, Dubiel excitedly called both PEMA and Dauphin County and asked both to have Maggie Reilly call him. She did, and then she called PEMA to recommend that Goldsboro and nearby Brunner Island begin preparations for a quick evacuation. Met Ed requested a helicopter from the state police to monitor the dosage readings in Goldsboro.

Since the Susquehanna River divided Dauphin and York counties, Colonel Henderson called York County Emergency Management Director Leslie Jackson. Jackson then began alerting local directors around Goldsboro to go on an evacuation alert.

With radiation at a then unconfirmed 10 rems per hour, immediate action was essential. Henderson called Governor Thornburgh, whom he had met only once and then briefly, to tell him about the preparations. Thornburgh asked Henderson to call William Scranton, the

Lieutenant Governor and chairman of the Emergency Management Council.[27] Thornburgh also asked his press secretary, Paul Critchlow, to contact Scranton and to find out what he could about the event.

About 8:15 A.M., the immediate crisis collapsed. Met Ed was finally able to fly radiation monitors to Goldsboro, and they discovered that readings were essentially at background levels. Dubiel informed Reilly, who informed Henderson, who informed both Scranton and York County. There was to be no panic evacuation. Now everyone could go back to the routine notifications as described in the emergency plan. Now there was time to consider in an orderly fashion whether the state was prepared for the emergency that was in progress.

Molloy in Dauphin County was convinced that it was not. In early 1975, he compiled the five mile radius plan inside Dauphin County for Three Mile Island.[28] Since that time, he had pushed local directors to improve the detail in their local plans. However, heading into this crisis, he did not feel that any of the local plans were adequate.[29] As a result, he spent much of the morning notifying the local directors and encouraging them to begin adding detail to their evacuation plans.

The emergency management structure in Pennsylvania had two characteristics that bothered Molloy before and during this crisis. First, the emergency management directors were appointed by the governor, but recommended at each level by the corresponding political leadership.[30] The governor appointed Oran Henderson; the county council recommended Kevin Molloy; the local mayors recommended the local directors. This system was designed to assure that the communications flowing through the emergency channels would be accessible to the local political leadership. It was assumed that the political leaders would recommend directors whom they could trust. The second difficulty for Molloy was that the local directors were volunteers who had other jobs.

Both of these characteristics made it difficult for Molloy and other county directors to order improvements on the local level. In 1978, a state law made the county directors responsible for training local directors.[31] There were several opportunities for this training. For instance, Met Ed held tests and gave instructions yearly at the plant for local emergency management personnel. However, the county directors could not force the locals to attend. Because some systematically avoided training, Molloy had some local directors whom he wanted to fire. However, he did not know how.[32]

Despite these problems, the evacuation machinery around the plant slowly and systematically began to swing into a state of readiness. Since there were no new crises evident at the plant, the locals kept informed, added details to their plans, and waited.

On the state level, however, developments were rapid. The Lieutenant Governor was notified of the emergency shortly after arriving at his office about 8:20 A.M. In addition to chairing the State Emergency Council, the Lieutenant Governor was also the head of the State Energy Council. In that capacity, he had a press conference on energy previously scheduled for 10:00 A.M.

It was soon clear that the regular energy press conference would have to wait. However, Dave Milne in public relations for the Department of Environmental Resources released the planned announcement that the press conference would be held. The press assumed that the conference would be about Three Mile Island.[33] Therefore, Scranton quickly found himself boxed into the role of government spokesman on Three Mile Island. Given his position on the State Emergency Council, this probably would have happened with time anyway.

But William Scranton knew next to nothing about Three Mile Island. The message he received from PEMA had been translated so many times that about all he was told was that the plant had "failed to fuel," an apparent reference to failed fuel. Fortunately, several energy staff people were due in at 9:30 A.M. to help with the final briefing for the regularly scheduled press conference. As they arrived, they would be able to help diagnose information.

Shortly after being notified, Scranton talked to Oran Henderson at length to find out what Henderson could tell him.[34] Unfortunately, Henderson was not an expert on these matters either. Henderson offered to attend the press conference to answer any questions on emergency preparedness. However, Scranton was concerned that Henderson's presence might distract from the purpose of the press conference and make things seem more serious than they were.[35] Therefore, Scranton declined the offer.

As 9:30 A.M. approached, the team in Scranton's office grew. Critchlow was there to report events back to the Governor. Joyce Freeman and Mark Knouse were on the Lieutenant Governor's staff. Dave Milne, Nat Goldhaber, and Robert Laughlin represented environmental affairs, energy, and the Governor's science advisory machinery.

But Scranton was still hopelessly unprepared to make a statement on the situation at Three Mile Island. At 9:30 A.M. he called the Bureau of Radiation Protection and asked them to send someone to brief him. Margaret Reilly and Director Tom Gerusky were busy on the phones, but William Dornsife was available. Dornsife called the plant for a detailed summary of the current state of affairs. Then he headed up the hill to the capitol to brief the Lieutenant Governor.[36] Scranton also reconsidered Oran Henderson's earlier offer. The PEMA Director was called again and asked to attend the briefing and press conference.

As 10:00 A.M. approached, the Lieutenant Governor clearly was not ready for the press conference. The briefing was completed, but it is crucial in such a delicate situation to issue a statement that is worded precisely and that is not likely to cause the public undue concern. Therefore, the press conference was delayed, and press anxiety grew.

Word that something was seriously wrong was spreading at amazing speed through the press. There were signs everywhere, such as the unmistakable attention surrounding the press conference preparations. The situation at the island seemed chaotic, with a restless mob of Met Ed employees milling around Route 441 trying to figure out why they were not being permitted to go to work.[37]

Meanwhile, as the Lieutenant Governor's group was preparing their carefully worded statement, events changed to make the statement no longer accurate. At 9:45 A.M., Dubiel at the plant called Margaret Reilly to report that off-site releases of radioactive iodine had been detected.[38] For a while, Met Ed and BRP discussed the mechanics of having their samples flown to Harrisburg for analysis; the Met Ed laboratory was contaminated. Arrangements were finally made for a state police helicopter to pick up the sample.

Just before the Lieutenant Governor went into his press conference, Tom Gerusky at BRP called Dornsife to tell him about the releases. The level was not very high, and they were later to analyze the samples and find that the level was lower than they thought.[39] Nevertheless, Dornsife now knew that there were off-site releases. Unfortunately, he did not have an opportunity to tell the Lieutenant Governor. The press conference had been delayed about an hour and could wait no longer.

As Scranton, Dornsife, and Henderson walked into the press room, the press was clearly aroused. Scranton read his general statement, saying that a turbine trip had caused a back-up in the system that spilled radioactive water.[40] There had been releases to the environment from the auxiliary building, but nothing had been detected off site. There was no need for evacuation, and the investigation was continuing.

The statement and the first few questions went well. But when Dornsife heard Scranton repeating the claim that no off-site releases had been detected, he felt he needed to interrupt. He cut off a reporter's question to say he had been informed that releases had been detected in Goldsboro.

The press conference immediately deteriorated. Dornsife noted that the Lieutenant Governor and his staff were displeased at the interrupting news.[41] The press also quickly became frustrated. As they hammered at Dornsife, they asked questions that had technical answers. Dornsife, who is a nuclear engineer, answered accordingly. Before long, no one in the room seemed to understand what had been said.

Scranton ended the news conference at 11:30 A.M. and went to brief the Governor. The first test of credibility with the press had gone poorly. But the situation would get much worse.

As Scranton left the Governor's office about noon, a reporter told him that Met Ed Vice-President John Herbein was at the plant and was saying that no off-site releases had been detected. Scranton had checked and verified Dornsife's findings and stood by the story that off-site releases of radioactive iodine had been detected. It did not sit well with Scranton that the plant seemed to be lying about the releases.

Scranton's patience was hit much harder just before 2:00 P.M. when Tom Gerusky called from the Bureau of Radiation Protection. BRP had been monitoring a continuous release from about 11:00 A.M. to about 1:30 P.M.[42] Just as the release was ending, the plant called BRP to report that it had been venting gasses.

Now Scranton was angry. Apart from apparently misrepresenting the condition of the plant, it now seemed that Met Ed intended to release radioactivity at will and without notifying the state. Scranton would not accept this. He called the plant and asked for a representative to attend a 2:30 P.M. meeting to explain what they were doing. Paul Critchlow was upset enough to want a lawyer present. Therefore, he requested that a deputy attorney general attend the meeting.[43]

Herbein, Kunder, and Miller showed up for the 2:30 P.M. meeting representing Met Ed. Herbein did most of the talking for Met Ed since he was the only representative from senior management. The other two were there to discuss technical issues if they arose.

As an exercise in communications, the meeting was a disaster. By the time it was over, Critchlow would not set up a joint news conference with Scranton and Herbein because he "wanted to preserve the Governor and the Lieutenant Governor's credibility."[44] From this meeting on, the state refused to associate itself with Met Ed.[45]

In the meeting, Herbein began by denying that anyone at Met Ed had said that there was an off-site release as a result of the accident.

Gerusky told him that the Bureau of Radiation Protection had monitored a release from 11:00 A.M. to 1:30 P.M. and that the state had
not been notified. Herbein then changed his story and admitted that
there had been such a release. However, he said that it was only
normal venting. The state officials were getting very upset.

Herbein said that there were still tanks that were overpressurized
and that more venting with releases might be needed later in the day.
He was asked why he did not say this during his own press conference earlier in the day, and he replied that "it didn't come up."[46] In
less than an hour, the meeting was ended. The state officials clearly
felt that it was serving no purpose.

Information was difficult to gather everywhere, except from the
news media. When private citizens rely on the news media for information, it is a normal state of affairs. But when government officials
do not know whether to believe official sources or media sources,
the official communications network is not doing its job.

According to the emergency plan, information to the local government officials was to be disseminated through each sector's emergency
management director. That information network was performing
reasonably well. PEMA asked the Bureau of Radiation Protection for
hourly updates from the plant. The information was then passed to
the local emergency management directors.

But through the afternoon, the only change in the plant was in the
slow, but growing, coolant flow through the hot legs. The appropriate message to be passed to the local directors was "no change."
Since the volunteer directors did not understand the situation, it
was not clear what was not changing.

Further, at this time of crisis, both the local directors and the
local political leaders seem to have largely forgotten about each
other. Mayors expected direct information from the Governor and
publicly brooded when they could not get it.[47] Some local directors
did not contact their counterpart mayors. The mayor of Harrisburg
found out about the crisis when he was called for an interview by a
Boston radio station.[48] But probably the best anecdote about the
breakdown was in the township of West Donegal, where Emergency
Management Director Jim Connors, a Three Mile Island employee,
was called in to work.[49]

Lieutenant Governor Scranton appreciated the critical need to
protect his credibility. Otherwise, people might not respond to state
directives during a time of real crisis. As a result, he scheduled a
4:30 P.M. press conference to report on his meeting with Met Ed. At
that press conference, reading from a prepared text, he stated that
the situation was more complex than the company had led everyone

to believe and that the company had given conflicting information. He said that there appeared to be no threat to the public, but the releases might not be over.

The press asked several questions for clarification. There was no doubt left by the answers that Scranton intended to give his own version and was not very interested in what Met Ed had said. Unfortunately, Scranton only vaguely understood the problem at the plant. While he was able to restore and/or protect his credibility, he did not have the words that would calm the anxieties of the press or the watching world. He had not forgotten that Herbein had warned that future releases might be required. While Scranton did not want to cause undue concern, he also did not want to get trapped between his desire to calm the public and the lack of credibility that occurs when optimistic predictions turn sour.

As the 4:30 P.M. press conference was announcing the "divorce" of Met Ed and the state government, the only continual communication network between the two sources was operating as strongly as ever. The Bureau of Radiation Protection received periodic updates from the control room. The updates were passed to PEMA, where they were disseminated to the affected counties. Both PEMA and BRP were, of course, on call at the Lieutenant Governor's request.

But the rules of the game were changing. As the Lieutenant Governor encountered increasing evidence that led him to believe that Met Ed was ignoring the state in their decisions, this crisis increasingly became his personal battle. Molloy in Dauphin County complained later that county efforts were crippled when the written plan was bypassed and all the decisions were discussed and made in closed meetings at the top.[50] By the end of the crisis, it was difficult to find decision-makers who were still following the plan. But for about half of Wednesday, it worked.

After the 4:30 P.M. press conference on Wednesday, however, the first open breach of the plan was obvious. Scranton wanted his own personal sources of information at the plant. He felt as if he had become the watchdog for the state, and he wanted to assess the competence and credibility of those in the control room. It is not clear whether it even occurred to him that he could get such information through PEMA or BRP. He wanted to see for himself.

At 5:30 P.M., Nat Goldhaber on the Lieutenant Governor's staff called James Higgins from NRC, who was now in the control room. Goldhaber was told about the progress in the hot legs and the decision to repressurize. The news was good, but very preliminary.

On three more occasions over the next two hours, different people from the Lieutenant Governor's staff called NRC employees in the plant. The information being passed was very helpful and rather optimistic. But Scranton had heard optimistic answers before, and he wanted to see the sources of the information in person.

At 7:30 P.M. Higgins, Neely, and Gallina were called and were asked to come brief the Lieutenant Governor. It was not a good time, since the plant was within minutes of trying to start a reactor coolant pump. However, there was a sizable contingent of NRC people in the control room, and these three were not needed. Therefore, they headed for the car.

Neely did not make it. In attempting to leave the plant, he set off the alarm indicating that he had radiation on his clothes.[51] Therefore, he was left behind, and Higgins and Gallina headed off toward Harrisburg to see the Lieutenant Governor.

Like all of the meetings and public contacts that day, the NRC meeting had its awkward moments. But on balance, it went well. Higgins and Gallina were joined by Robert Friess of the DOE monitoring team. State officers included Scranton, Critchlow, Knouse, and Gerusky. Scranton asked State Representative Deweese, Colonel Henderson, and his assistant Williamson to wait outside at first because the room was too crowded.[52] Representative Deweese became offended by this and complained publicly.[53] However, Deweese, Henderson, and Williamson were later admitted. So was Jay Waldman, representing the Governor.[54]

Throughout the meeting, there was a continual problem that Higgins and Gallina gave a very technical summary to an untrained audience. When the state people tried to force the summary down into common terms by asking questions like "What the hell happened here," the NRC people did not know how to respond well.

The meeting lasted about two hours with interruptions. In that time, it was finally hammered out that there was a slight chance of a meltdown, but that there would be twenty to thirty hours notice if that happened. Having agreed to that much, Scranton then arranged to have the NRC people visit the Governor's mansion later that night.

First, Scranton invited the two to a 10:00 P.M. press conference, at which Scranton announced that radioactive material had escaped to the auxiliary building where it was being slowly vented into the atmosphere. It was causing off-site releases, but not enough to cause concern. Scranton then invited the press to question Higgins and Gallina.

By hindsight, Higgins and Gallina gave the press an unjustifiably optimistic review. They claimed that there was no substantial damage to the plant or the core. They said that the auxiliary building water could be cleaned up in about a day and that the plant should be in a state of cold shutdown by then. The news was received reasonably well, and the conference ended.

Scranton now escorted Higgins and Gallina to the Governor's mansion. Here they replayed the awkward sequence of technicians and politicians trying to speak the same language. But Thornburgh was accomplished at cross-examination and soon arrived at the same conclusion about the slim possibility of a meltdown that the Lieutenant Governor had reached. With that, the briefing was over, and Higgins and Gallina returned to the plant.

By late Wednesday evening, the plant appeared stable. Reactor coolant pump 1A was maintaining a strong flow through the loop. Miller was technically in charge until he left at 3:00 A.M., although senior management was now checking all decisions.[55] The big effort now was the attempts to stop the releases from the auxiliary building. The complicated vent header was being troubleshot to try to find its leak, although there was no way in all that radiation to fix it. Slabs of a substance called poly were laid over the radioactive water to slow its evaporation. Despite these efforts, the highest off-site releases of the day were recorded at 10:30 P.M.

The emergency management machinery remained on alert, although the crisis seemed abated. There had been a substantial number of voluntary evacuations, particularly in Goldsboro. Within a day, these voluntary evacuations would reach 40 percent of the population and would be dearly appreciated by those planning to evacuate the rest.

The state political decision-makers finally went home to go to sleep. Years before, the Governor had read the book *We Almost Lost Detroit*, a dramatic exposé that purported to tell the story of an earlier accident at the Fermi experimental reactor. He remembered the ghastly description in the book of the problem of core damage.[56] As he waited to go to sleep, he recounted the meeting with Higgins and Gallina. He wondered why they had breezed so lightly over the issue of damage to the core.

NOTES

1. "Report of the Office of the Chief Counsel on Emergency Preparedness" to the President's Commission on the Accident at Three Mile Island, October 1979, p. 1; see also deposition of Harold Collins to President's Commission, Washington, D.C., July 28, 1979, p. 18.

2. Government Operations Committee, "Emergency Planning Around U.S. Nuclear Powerplants: Nuclear Regulatory Commission Oversight," Report 96–413, U.S. House of Representatives, 96th Cong., 1st sess., August 8, 1979, p. 3.

3. "Demographic Statistics Pertaining to Nuclear Power Reactor Sites," NUREG 0348 (Washington, D.C.: Nuclear Regulatory Commission, December 1977), table one.

4. Deposition of Robert Ryan, President's Commission, Bethesda, Maryland, August 2, 1979, pp. 71-72.

5. NRC Regulatory Guide 1.101.

6. Deposition of Harold Collins for President's Commission, p. 7.

7. Deposition of Robert Ryan for President's Commission, pp. 33-34.

8. Ibid., p. 27.

9. "Report of the Office of the Chief Counsel on Emergency Preparedness," p. 9.

10. U.S. General Accounting Office, "Report to the Congress: Areas Around Nuclear Facilities Should Be Better Prepared for Radiological Emergencies" (Washington, D.C.: U.S. Government Printing Office, March 30, 1979), p. 1.

11. Testimony of Colonel Oran Henderson, President's Commission, August 2, 1979, p. 33.

12. Personal interview with Joyce Freeman, Harrisburg, Pennsylvania, September 14, 1979.

13. Scenario taken from "Report of the Office of the Chief Counsel on Emergency Preparedness," pp. 11-13.

14. Testimony of A.S. Kinsinger, Pennsylvania Select Committee on Three Mile Island, August 7, 1979, p. 16.

15. Written submission of Gary Miller, *Accident at the Three Mile Island, Part II*, p. 263. See also, "Report of the Office of Chief Counsel on Emergency Response" to President's Commission, October 1979, p. 2.

16. Deposition of William Dornsife for President's Commission, Harrisburg, Pennsylvania, July 24, 1979, pp. 17-18.

17. Testimony of Paul Leese, Director of Emergency Management for Lancaster County, Pennsylvania Select Committee on Three Mile Island, July 24, 1979, pp. 22-23.

18. Testimony of Aubrey Godwin, *Brown's Ferry Nuclear Plant Fire*, hearings before the Joint Committee on Atomic Energy, U.S. Congress, 94th Cong., 1st sess., September 16, 1975, pt. 1, p. 176.

19. Deposition of William Dornsife for President's Commission, pp. 17-21.

20. Scenario from "Report of Chief Counsel on Emergency Response," p. 3.

21. Deposition of William Dornsife for President's Commission, pp. 18-20.

22. Ibid., p. 20.

23. Testimony of Kevin Molloy, President's Commission, August 2, 1979, pp. 5-6.

24. "Report of the Office of Chief Counsel on Emergency Response," p. 4.

25. Testimony of Major General Richard M. Scott, Pennsylvania Select Committee on Three Mile Island, August 8, 1979, p. 53.

26. Deposition of Tom Gerusky for President's Commission, Harrisburg, Pennsylvania, July 24, 1979, p. 31.

27. Deposition of Richard Thornburgh for President's Commission, Harrisburg, Pennsylvania, August 17, 1979, pp. 5-6.

28. Testimony of Kevin Molloy, Pennsylvania Select Committee on Three Mile Island, July 25, 1979, p. 26.

29. Testimony of Kevin Molloy, President's Commission, August 2, 1979, pp. 3–4.

30. Testimony of Kevin Molloy, Pennsylvania Select Committee on Three Mile Island, July 25, 1979, p. 36.

31. Testimony of Paul Leese, Pennsylvania Select Committee on Three Mile Island, July 24, 1979, p. 22.

32. Testimony of Kevin Molloy, Pennsylvania Select Committee on Three Mile Island, July 25, 1979, p. 35.

33. Personal interview with Joyce Freeman.

34. Unless otherwise stated, information from personal interview with Joyce Freeman.

35. "Report of the Office of Chief Counsel on Emergency Response," p. 12.

36. Deposition of William Dornsife for President's Commission, pp. 25–27.

37. "Report of the Office of Chief Counsel on Emergency Response," p. 11.

38. Deposition of Tom Gerusky for President's Commission, p. 34.

39. Deposition of William Dornsife for President's Commission, p. 29.

40. Summary from testimony of William Scranton, President's Commission, August 2, 1979, p. 183.

41. "Report of the Office of Chief Counsel on Emergency Response," p. 13.

42. Testimony of Richard Thornburgh, Pennsylvania Select Committee on Three Mile Island, May 10, 1979, p. 11.

43. "Report of the Office of Chief Counsel on Emergency Response," p. 18.

44. Ibid., p. 19.

45. Testimony of Richard Thornburgh, President's Commission, August 21, 1979, p. 58.

46. Report of the Office of Chief Counsel on Emergency Response," pp. 18–19.

47. Testimony of Mayor Albert Wohlsen of Lancaster, Pennsylvania Select Committee on Three Mile Island, July 24, 1979, p. 81.

48. Testimony of Paul Doutrich, President's Commission, May 19, 1979, p. 125.

49. Testimony of Paul Leese, Pennsylvania Select Committee on Three Mile Island, July 24, 1979.

50. Testimony of Kevin Molloy, Pennsylvania Select Committee on Three Mile Island, July 25, 1979, p. 8.

51. Deposition of James Higgins for President's Commission, King of Prussia, Pennsylvania, August 17, 1979, p. 31.

52. Testimony of William Scranton, Pennsylvania Select Committee on Three Mile Island, May 10, 1979, p. 52.

53. Testimony of William Deweese, Pennsylvania Select Committee on Three Mile Island, May 10, 1979, p. 21.

54. Meeting summary from "Report of the Office of Chief Counsel on Emergency Response," pp. 22–23.

55. Written submission of Gary Miller, *Accident at the Three Mile Island, Part II*, p. 277.

56. Deposition of Richard Thornburgh to President's Commission, p. 25.

Chapter 8

Too Many Heroes and No Toilets

Like almost everything attached to the Three Mile Island incident, the state emergency plan has been subjected to its share of attacks. The NRC did not concur in it. The Administrator of the Federal Disaster Assistance Administration has called it "so generalized as to be virtually worthless. . . ."[1] When in use, it certainly had moments of chaos.

But "chaos" was a term waiting for definition until the federal agencies arrived. By mid-Wednesday, no one any longer knew how many federal agencies were involved or collecting useful information on the Three Mile Island accident. By Friday morning, the federal actors in charge helped turn a crisis into a near calamity. By Monday, the federal coordinating effort escalated into a war of political egos and a media sideshow.

Even on paper, the federal response mechanism was a mess. To begin, there was no one response plan. There were three or four, depending on definitions. Each plan involved different agencies in different relationships. Each was written to the apparent exclusion of the others. None was apparently taken very seriously because federal agencies almost immediately began to intervene without invitation or authorization.

The oldest plan was the Interagency Radiological Assistance Plan, or IRAP, dating from 1961. Before 1961, no federal agency had the responsibility to plan for or to respond to a peacetime radiological incident. Four years before, there had been a major reactor accident involving international radioactive contamination in Windscale, England. The only U.S. release was in 1961 at a military reactor in

Idaho and had been contained on the military base. Still, proceeding without a plan seemed to be a major mistake.

The IRAP was a memorandum of understanding. Every agency with some potential to respond to a radiological emergency agreed that its efforts would now be coordinated by the Atomic Energy Commission. The IRAP did not add any new monitoring capabili-ties. It was simply an agreement on coordination. But the coordina-tion was crucial, since the AEC, and then the NRC, had never attempted to maintain the most sophisticated radiological-monitoring equipment available.[2]

In 1975, the AEC split into ERDA and the NRC. It was not clear who inherited the coordination rights under IRAP, so negotiations were started. It took two years for the two new agencies to reach a new agreement on planning, preparedness, and response to emer-gencies, which was finally written and signed in 1977.

But the agreement still did not say who was in charge. It agreed that the IRAP still existed and that it was the vehicle by which NRC could get IRAP support. ERDA, which has now been moved into the Department of Energy (DOE), was specified as the lead agency once the plan was operating. But it was not clear whether DOE or NRC initiated the plan and requested assistance.[3]

It is also not clear whether IRAP was actually used during the Three Mile Island response. It was definitely used as a justification for some of the intervention, but not in such a way that any IRAP coordination took place. Rather, the notification process in IRAP allowed each agency to make its own attempts to intervene uni-laterally with the state.

At 9:05 A.M., the NRC notified the Environmental Protection Agency's (EPA) Office of Radiological Programs of the event in prog-ress.[4] EPA then established its own three man command center con-sisting of Administrator Costle, his chief science advisor Richard Dowd, and Assistant Administrator Stephen Gage. They called Maggie Reilly in Pennsylvania to offer assistance. At the moment, the utility and the Bureau of Radiation Protection had monitoring in progress, and NRC was on the way. Ms. Reilly thanked them for the offer, but she did not need the assistance. EPA then remained dor-mant until the crisis flared up on Friday, when they and almost everyone else moved in without invitation or opposition.

When the Department of Energy heard of the crisis on the morn-ing of March 28, they called Lieutenant Governor Scranton to ask permission to monitor. Not wanting to turn down free help, he invited them in.[5] Within hours, a DOE monitoring team was flown in from Andrews Air Force Base and set up headquarters in a trailer at Harrisburg's Capitol City Airport.[6] Very quickly, they accumulated more off-site data than everyone else involved.

But almost no one knew that they were there. Dubiel at the plant, who was coordinating radiological data with NRC's Gallina on off-site releases, never heard from them.[7] The relationship of the DOE group to the NRC is less clear. The NRC insists that it did not know of the DOE team until days later. But the radiation-monitoring teams in the plant moved to a trailer on the observation center lawn Wednesday evening. Joe Deal from the DOE team was also in the trailer. However, both the state and the NRC representatives in the trailer failed to realize that Deal had his own team at work.[8]

It was also unclear who the Department of Energy team represented. They repeatedly used the title of IRAP, indicating that they were the coordinating body. However, they decided early Wednesday that they did not need to ask any other IRAP signatories to join them.[9] Also, Deal was not sure in his role as the on-site coordinator whether he was supposed to be answering to the NRC.[10] Instead, the DOE team lived in something of a twilight zone at the Capitol City Airport, gathering information that was not getting to people who could use it. But DOE did not realize that there was a problem. They held daily press briefings with handouts, and they assumed that NRC must be getting copies.[11]

IRAP is not the only federal coordinating plan, and some agencies did not accept that it was still in force in 1979. In 1974, the IRAP concept was amended by the Disaster Relief Act, which permitted the Federal Disaster Assistance Administration (FDAA) to assist state and local governments during declared disasters. While leadership of IRAP was soon to be fought out between ERDA and NRC, FDAA was also given a strong role in IRAP by the Disaster Relief Act.

But the Federal Preparedness Agency (FPA), which is normally in charge of wartime planning, made what could easily be described as a power play. In 1974, they began circulating drafts of a "Federal Response Plan for Peace-time Nuclear Emergencies" to thirty-two federal departments and agencies. It divided nuclear incidents into four categories, with different agencies responsible for each. The incident at Three Mile Island threatened to grow into category 3, which was assigned to FDAA.

By December 1976, all of the thirty-two agencies had agreed to the category 3 plan except FDAA, who saw the entire scheme as an attempt by an unauthorized agency to limit FDAA's jurisdiction. To get past this roadblock, FPA issued an "Interim Guidance," which allowed them to continue to negotiate with FDAA while FPA started work on categories 1, 2, and 4.

During the summer of 1977, FDAA and FPA finally agreed to the wording for category 3. However, FDAA first had to get permission from its parent Department of Housing and Urban Development. Donald Carbone on the FDAA staff sent a copy for concurrence to HUD on September 26, 1977.[12] It was almost a year before HUD responded to say that they needed another copy since they had lost the first one. A second copy was quickly sent, and HUD approved the new wording. HUD notified FPA by letter of its acceptance. FPA promptly lost that letter by routing it to an individual who had changed jobs.

But there were people in FPA who were still working on this plan, and they called FDAA periodically for months. Apparently, only Carbone in FDAA was aware that the wording had been approved and forwarded to FPA. From October 1978 until after the Three Mile Island accident, FPA was unable to find the person who could tell them that the plan had been approved, but that they had lost the acceptance letter.

The consequence of this confusion going into Three Mile Island was that neither FPA nor FDAA knew whether IRAP was still in effect. But they were ahead of most agencies. The political leadership of EPA and HEW, who committed their agency resources later in the crisis, were not aware there was an IRAP or that their agencies were members.[13] Neither was Jack Watson, who headed the White House's interagency task force on emergency planning.[14]

There were other emergency plans available involving single federal agencies. The Department of Energy, which early in the accident claimed the rights to the IRAP, also had a second plan. Long ago, the Atomic Energy Commission established a radiological assistance program to coordinate the use of national laboratories and monitoring equipment during an emergency. That plan was now in the Department of Energy.

As part of TMI's notification procedure, they called the administrators of this program at the Brookhaven National Laboratory. Brookhaven was alerted to the site emergency at 7:09 A.M. and to the general emergency at 7:30 A.M.[15] However, Met Ed did not require assistance.

Shortly after 8:30 A.M., Brookhaven called Margaret Reilly to see if the state needed assistance. She declined the offer at that point, but said that she would call back later in the day.[16] Brookhaven then alerted DOE's Emergency Action Coordinating Team at Germantown, Maryland, and they stayed on alert.

At 9:45 A.M. Brookhaven called Ms. Reilly back. By this time she suspected that off-site releases were occurring. Therefore, she accepted the offer.[17] DOE dispatched a team by helicopter from Brookhaven, and they were soon working in coordination with the Bureau of Radiation Protection.

But not all assistance was invited or welcome. PEMA notified the Defense Civil Preparedness Agency regional office in Olney, Maryland, at 8:45 A.M.[18] However, there was nothing that they could do to help at this point other than be on alert. It so happens that the Defense Civil Preparedness Agency funds about half of PEMA's budget. When Assistant Director John McConnell heard of the accident later in the morning, he ordered a field representative from Olney to go to PEMA and stay there until further notice. Oran Henderson never requested this observer, although he was in no position to object. Therefore, the observer went to Harrisburg and "observed." In later explaining his actions, it did not much interest McConnell that the intrusion was carried out without invitation or even legal authorization. He felt that he had the right to watch his money at work.[19]

FDAA Administrator William Wilcox learned of the accident through a news service bulletin at 10:00 A.M.[20] He called his regional director, Robert Adamcik, to suggest that Adamcik go to PEMA as an observer. Wilcox was convinced that an emergency would soon be declared and that FDAA would be coordinating all federal efforts.[21]

Adamcik had not heard of a problem at Three Mile Island and objected. When the regional office called PEMA and was told that they were not needed, Adamcik did not go to Harrisburg. For a while, he got away with it.

But the plan that really mattered at first at Three Mile Island was the one that grew out of the NRC, PEMA, the Bureau of Radiation Protection, and the Lieutenant Governor working together. It was not the plan that had been drawn up in advance. By Wednesday evening, the county and local emergency management directors were receiving nothing more than polite notification.

It was not a plan glued together with trust. While PEMA, the Bureau of Radiation Protection, and the NRC had a good working relationship on the phone, the Lieutenant Governor's office insisted on treating all sources at the plant with a degree of unmistakable caution. The cautious acceptance held through most of Thursday, although it had to be reinforced several times.

Finally, it was not a plan that was even vaguely prepared to evacuate anyone. As the state later learned, evacuation required heavy local participation. But local participation in decision-making basically disappeared by Thursday. While the locals continued to work on their evacuation plans on Thursday, they lost the urgency that they had felt on Wednesday.[22] Fortunately, the reactor gave almost everyone this day of rest.

In the reactor control room, early Thursday was a confusing time. There was an intense desire to believe that the difficulties in the core were over. Both the primary and the secondary loops were flowing, although through amended paths. The operators were certain that the reactor core was covered, and they were correct.

However, they had too many readings that did not make sense. Until they could sort out what had happened on Wednesday, they could not be certain about the security that they wanted to feel on Thursday.

There was plenty of help available. At the observation center a few hundred yards away, the number of trailers was growing as nuclear experts from throughout the country gathered at Met Ed's request to help with the analysis. This group was labeled the "Industry Advisory Group," and the expertise that it represented was impressive.[23] This group of experts was supplemented by several hundred General Public Utilities employees who were assigned to the plant rather than to their regular jobs.

Equally important, the Incident Response Action Center in Bethesda was alive with activity. Harold Denton finally went to the center about 5:00 P.M. on Wednesday. He spent the night, but had to leave in the morning to testify before Congress.[24]

But most of the high level administrators at NRC were there and working on questions that fit their expertise. For instance, there was the confusing matter of the core thermocouples. Most were still off-scale and presumed shorted out. But some of them were occasionally coming back on-scale, and one was actually rising.[25] Shorted thermocouples should not act that way.

Roger Mattson in Bethesda decided to research the situation. During the morning on Thursday, he was able to find out that the thermocouple probes should not short out from heat until they reach 2300–2500°F. If the core had reached those temperatures, there was much more involved in those first few hours than anyone then imagined. But if the thermocouples had not shorted, they were now clearly indicating that the plant was in serious trouble. This problem needed more thought.

There were other points of confusion. In fact, almost everything that had occurred on Wednesday was confusing. Therefore, the NRC and Met Ed employees spent much of the day going back through the various indicators and strip charts, trying to retrace what had happened.

But one big piece of information was missing. If the operators could just find out how significant the fuel damage had been, they would be much closer to understanding what had happened and what the core was doing now. There was a way to find out. If they could draw a sample of core-cooling water and analyze its contents, they would know how much fuel cladding had failed.

But drawing the sample was tricky in a radioactive lab, and the analysis was lengthy. They could not have readings on a core sample until that evening. Therefore, for now, they attended to the more pressing matters.

The biggest pressing matter was radioactivity. Hundreds of thousands of gallons of radioactive water were standing in the basement of the containment building and on the floor of the auxiliary building. In the auxiliary building, it was still causing vented releases to the atmosphere.

The off-site releases were down from the high of 10:30 P.M. the evening before. However, on-site readings at the auxiliary building vents varied tremendously from quite low to a recorded high of 3000 millirems.[26] There was no alarm at the plant. Water was being pumped off the floor and into storage tanks, and it was stirring up temporary radioactive puffs.[27] Over time, the puffs were decreasing. Unfortunately, the plant would have to find a new place to put some of its water.

Outside of the plant, Three Mile Island had become a media extravaganza. The press were everywhere, and special security arrangements had to be made to keep them out of the back lawn of the observation center, where "trailer city" was quickly taking form.

But the utility was not trying to maintain a low profile. Med Ed President Walter Creitz appeared on the "Today Show." Senator Gary Hart was also on the program and announced that he and other members of Congress planned to visit the site at noon.[28] Reporters searched the streets of Goldsboro and Middletown for citizen reactions and the human interest stories that fill in between major news developments.

Ironically, about the only place not swamped with reporters on Thursday was the state emergency machinery. Within a day, it would be difficult to get in or out of PEMA headquarters because of all the reporters. But on Thursday, the staged events at the plant were giving the officials in Harrisburg some much-needed time to plan in peace.[29]

Most state officials had a late evening on Wednesday, and sleep did not come easily. Nevertheless, most of the principal actors were back on the job at about the normal time the next morning. Shortly after arriving, Scranton and Radiation Protection Director Tom Gerusky agreed that they would be more comfortable with a personal representative on site. They selected William Dornsife, who was an obvious choice since he was a nuclear engineer.

Dornsife arrived at midmorning and quickly began sending back information. His basic information was the same as what the NRC had been saying. The control room was calm, and there was a sense that the reactor problems were over. The mopping up task was difficult and led to some disturbing readings. However, everyone expected the releases to decline shortly.[30]

Scranton was pleased with this report, but he had become distrustful the day before. During the morning hours, he came to the conclusion that he was not going to be at ease with the answer that the crisis was over until he saw the plant himself. He discussed the idea with Governor Thornburgh, who agreed that a visit to the plant would be a good idea.[31] Therefore, shortly before 10:00 A.M., Scranton asked some of his assistants to arrange for him to tour the plant.

Several staff people began working on the tour. But at this point, Met Ed senior management was not in any mood to do Scranton any favors, even if they had to be careful how they responded to him. Besides, the radiation problems in the plant were intense, and the control room was very crowded. Therefore, Scranton's staff worked for over an hour trying to arrange the tour.[32]

Finally, Scranton called Met Ed President Walter Creitz personally and asked to go on a tour of the plant. Creitz was very reluctant to allow him to go.[33] However, he finally consented. Creitz pointed out that the senators would be there at noon, which would be about the time that Scranton would arrive. Creitz suggested that it might be best if they toured together.

But Scranton had some specific questions he wanted to ask, and he especially wanted to see how much radioactive water was still loose in the auxiliary building. Scranton did not know what the senators wanted to see, and he did not want to be constrained by their schedules or priorities.

It was finally agreed that Senators Hart and Heinz would be kept at the observation center while Scranton went through the plant. Scranton arrived at the plant about noon and spent over three hours touring the facility. The tour was often interrupted by floating radiation and the need to don protective clothing.

However, Scranton verified for himself what he had been told. The control room atmosphere was busy but calm. There was a considerable amount of cleaning up to do, and the radiation in the auxiliary building was about 3000 millirems per hour.[34] Scranton left about 3:30 P.M. and reported these findings to the Governor and then to the world in a 5:15 P.M. press conference.

But when Scranton returned to the office, there were four new disturbing developments that hit him in rapid succession and made him an unquestioned pessimist by the end of the evening. First, another uninvited federal agency was creating havoc within the state bureaucracy.

Gordon MacLeod was sworn in as Pennsylvania Secretary of Health on March 16, twelve days before the accident. In that period, he had found the time to learn the rudiments of his job and to appoint a few assistants. But it quickly became clear to him that he had a considerable amount left to learn.

On Wednesday morning, the Director of Health Communications at the central office in Harrisburg telephoned him in Pittsburgh, where he was visiting a branch office.[35] MacLeod asked to be put in touch with the person in charge of radiation health within the Department of Health. The Director of Communications had to explain that that office was in the Department of Environmental Resources. MacLeod then asked to be put in touch with the department's liaison to the Department of Environmental Resources. He was told that there was no liaison.

He finally asked the Communications Director to go to the departmental library and gather together information on radiation health for MacLeod to read. He was informed that the department had no library. It had been abolished two years earlier as a budgetary move. At this point, MacLeod realized that Three Mile Island was not intended to be his concern, and he went back to his normal business.

He heard no more about the problem on an official level until he was back in his office Thursday afternoon. While Scranton was touring the plant, MacLeod received a call from Dr. Anthony Robbins, who was Director of the National Institute of Occupational Safety and Health within the federal Department of Health, Education, and Welfare. The two had known each other slightly for years.

Robbins told MacLeod that he was very concerned about the con-dition of the reactor. MacLeod assured him that the radiation levels off site were quite low. Robbins said that he was worried about the state of the reactor, not about the releases. He claimed that the plant was in an "experimental mode" and that the operators were not sure how to shut it down in this condition.

At this point, the two participants vary widely on the content of the rest of their conversation. Robbins claims that the call was an informal contact to offer support and assistance.[36] Robbins is sup-ported in this claim by John Villforth at the Food and Drug Admin-istration in HEW.[37] Villforth and Robbins had earlier agreed that Robbins should call MacLeod and offer federal support and advice. However, Villforth was not a party to the actual call.

According to MacLeod's version, Robbins said he had been in con-tact with the Bureau of Radiation Health within the federal Food and Drug Administration. He said that based on his experience and the consultation, he was urging an evacuation.[38]

Whatever was actually said in that conversation, MacLeod felt that he had been strongly urged to recommend an evacuation. He was not sure that an evacuation was needed or wise, but he did not want to simply ignore the message. Therefore, he arranged a conference call among himself, Colonel Henderson at PEMA, Tom Gerusky at the Bureau of Radiation Protection, and John Pierce on Scranton's staff. He told them what Robbins had said. They agreed that there was no reason to act now. However, they also agreed that they would reconsider their options if they had any indication that the plant was in this "experimental mode."

Then MacLeod sprung a new idea on them. As a physician, he was aware that radiation has a much more significant impact on fetuses and developing children than on adults. He suggested that the Gover-nor might want to consider advising children under the age of two and pregnant women to voluntarily evacuate the area.[39] No one thought that this was advisable at that point. However, the idea of an advisory evacuation for the young had now been introduced into state governmental circles.

Most of this transpired while Scranton was at the plant and then on his way to brief the Governor at 3:45 P.M. Scranton did not hear about the evacuation advisory until he returned from the Governor's office. While at the Governor's office Scranton's trust of the federal government eroded further.

With Scranton at the plant, the staff people working on the issue gravitated toward the Governor's office. With the staff people milling around his office, there was a greater tendency for Thornburgh to get involved in the minute-by-minute operations. However, there had been some recent scares that had the public aroused. Rumors of evacuations in progress were rampant. To complicate matters, Dr. Ernest Sternglass of the University of Pittsburgh was interviewed during the afternoon on a Harrisburg radio station, and he recommended an evacuation of children and pregnant women.

Thornburgh finally decided that he needed to make a public statement to allay the citizens' fears. He scheduled a press conference for 4:00 P.M. although it was repeatedly delayed and finally began at 5:15 P.M. In order to prepare, he asked for Higgins, Gallina, and Scranton to meet with him at 3:45 P.M. for a briefing.[40] During that meeting, each gave his impressions that the plant was calm and that the crisis was receding. Each felt that the crew in the control room seemed competent.

Thornburgh asked about his concern that there was obviously some core damage. The participants relayed the common assumption in the control room that while some core damage had occurred, conditions were now stabilizing. With that, it was time to go to the press conference.

There are certain people who, during a crisis, have the demeanor to deliver even bad news in a way that calms people down. Harold Denton was later to be credited with such a talent. But it is also a fair description of Richard Thornburgh. Thornburgh's message was quite optimistic. He noted that conflicting stories were circulating in the media and the rumor mill. He emphasized that he had spent the last thirty-six hours listening to all sources before reaching the conclusion that there was absolutely no cause for alarm. He asked people not to act in haste.

The rest of the news conference followed this theme, but Scranton and Thornburgh felt that it finally drifted over the line of truthfulness. Charles Gallina was the NRC expert on site on radiological issues. Three different times during the questioning, he stated that the danger from off-site releases was over.

Scranton and Thornburgh knew that that was not true. While the reactor might be stable, both knew that there was still water to clean up in the auxiliary building and that releases would continue while the cleaning was underway. Gallina apparently still believed his optimistic predictions that the job could be finished on Thursday. But Scranton and Thornburgh both felt that Gallina was being dishonest.[41]

Not long after the press conference was over, Scranton heard about the strange experience of Gordon MacLeod. His suspicion was growing, although he still had nothing firm to document his distrust. But the first two events were insignificant compared to the third.

Every nuclear powerplant has discharge permits and limits for water and gas releases from the plant. Releases occur from numerous sources, including air-conditioning vents, drains, and flushing toilets. Like every discharge permit, there are limits put on the toxicity and oxygen depletion ability of the discharge. But nuclear powerplants, for obvious reasons, also have strict limits on the amount of radiation that they may release.

Water discharges from the Three Mile Island plant were first collected in waste tanks. At appropriate intervals, the contents were diluted and filtered to meet pollution control requirements and released into the Susquehanna River. This was the same type of waste treated to the same levels of cleanliness as was discharged from several sewage treatment plants along the river.

But radiation was a different matter. Early Wednesday morning, the auxiliary operators monitoring this system were concerned that some radioactive gasses might be suspended in the water. Their concern turned out to be well founded. Not taking any chances, all filtering and discharging was stopped.

But waste water collects quickly. By 2:30 P.M. Thursday afternoon, over 400,000 gallons of mildly radioactive waste water had been collected in the waste tanks. This was in addition to a roughly equal amount of highly radioactive water in containment and the auxiliary building.

Reactors are built with many capabilities, but they cannot store infinite amounts of waste water. By noon on Thursday, with the core and the radioactivity shoved into the background, the crisis of Three Mile Island was sewage. If they did not filter and discharge it quickly, it would literally overflow into the turbine building drains. Within minutes, it would enter the Susquehanna River undiluted.

Met Ed proceeded as carefully as they could. They obtained readings to indicate that the water met both water and gas release standards.[42] They collected gas data since the radioactive gasses in the water would supposedly separate in the river.

Then they approached Charles Gallina as the ranking NRC spokesman for permission. He had no objections if the discharge met the legal limits. To double check, Met Ed called NRC's Region I and talked to George Smith, the Chief of Fuel Facilities and the Safety Branch. Smith did not object. Before releasing it, Dubiel asked the NRC mobile lab to test a sample of the water. They did and found it to be well within limits.

By hindsight, Met Ed's appreciation for the political value of asking permission was amazing, considering that they had the legal right to dump the water anyway. Nevertheless, Dubiel also called Maggie Reilly and Tom Gerusky at the state before proceeding. They did not object either.

About 2:45 P.M., Dubiel began dumping the water. But even before he started, communications about the discharge began to get snarled. Gerusky called Region I to say that he thought NRC should give permission on this.[43] The message was passed to George Smith, who called Bethesda to get their permission. Lee Gossick and John Davis on the emergency management team both approved the dumping.

For the next few hours, the topic continued to pop up in conversations within or involving the center in Bethesda. After three hours, at 5:45 P.M., NRC Commissioner Hendrie heard about it, although the details he had were sparce. Hendrie and Commissioner Aherne called Edson Case and John Davis at the center to find out what was happening. The commissioners were told that the level had acceptable limits of radiation and that the state and Bethesda had been notified.

But Hendrie was quite anxious about the subject. He ordered the dumping stopped. He was concerned that this could cause the NRC embarrassment with the press, the state, and the federal government. His fears proved to be well founded.

During their conversation, Case passed the word to Region I to have the discharge stopped. George Smith, who had approved the original discharge, tried to reach Hendrie to explain the necessity of the discharge. But in his own words, he "got hell for asking questions."[44] Therefore, Region I Director Boyce Grier passed the order to Gary Miller at the plant. Miller was baffled, but he had to comply.

An amazing new complication was introduced into the recovery process for the reactor. For the next several hours, no one on the island could use a toilet. But the toilet crisis was just beginning and was to grow into a tremendous clashing of wills over the next few hours. Margaret Reilly, who was caught in the middle of it, was later to comment: "I thought, boy, if it were a *real* problem, what do you do then?"[45]

The 6:00 P.M. decision to stop the discharge caught the Bureau of Radiation Protection by surprise. Reilly and Gerusky thought the decision was a minor administrative detail, and they had not passed any notification to the Governor's advisors. Instead, Dave Milne in the Governor's office heard of it a little after 6:00 P.M. when a reporter called him from Washington to find out about an alleged dumping of radioactive water.

Now the Governor's office began its own investigation on the suspicion that someone was dumping "radioactive" water into the river. The water was mildly radioactive, and the confusion between the waste water and the auxiliary building water also arose that evening in Bethesda and in the press. However, when Milne called Reilly, he found out what had occurred.

Milne then told the Governor's press secretary, Paul Critchlow, who told the Governor.[46] But the message had begun losing important details by the time it reached the Governor, and there were things that the Governor wanted to know. For instance, he wanted to know how much radiation was involved and whether the water really needed to be dumped.[47] Milne checked with Bethesda and returned with the answers.

But the toilet crisis broke into open warfare between the state and the NRC over the next issue. The NRC had stopped the discharge at 6:00 P.M. Now someone had to take the politically dangerous step of ordering it restarted. The authority was NRC's. But the NRC was afraid of the implications and tried by an amazing set of fraudulent statements to get the state to accept the blame.

At 8:00 P.M., Milne called Jim Sniezek at Bethesda to get some answers for the Governor. Milne was told that the discharge had been stopped to allow the state time to notify users downstream. Milne was left with the impression that the Governor could restart the dumping when he felt ready.[48] Milne and his assistant, Peter Duncan, began writing an appropriate press release to have the Department of Environmental Resources restart the dumping. But as they were writing, they began to realize that they had never actually been told that the state had the authority to start the dumping. NRC had a press officer, Karl Abraham, in an office next to Paul Critchlow's. They went to him to get a clarification. That was a mistake.

Abraham tried to protect what he saw as the best interests of his home agency, the NRC. If he could convince the state to take the blame, he would. Some of the tactics he used over the next three hours of negotiations appalled the state decision-makers and drove a huge rift between the state and the NRC.

Abraham was not giving what Milne and Duncan considered to be straight answers. Instead, Abraham constantly shifted questions back over the phone to Bethesda. From that end, he was getting support. As NRC Commissioner Hendrie said during the debate "the Governor himself is going to make this decision. Let the Governor decide whether we are going to dump the crappers in the water."[49]

On the state side, the negotiators proliferated as it became clearer that the NRC was stalling. Paul Critchlow and Scranton's aide, Mark Knouse, joined the discussion. Finally, a little before 10:00 P.M., they got Abraham to admit that NRC was actually responsible for the decision.

Critchlow told him that the Governor did not mind making this decision. However, if it was NRC's decision to make, then the Governor was not going to take the political heat. Abraham responded with the amazing rejoinder, "Well, you have got a Lieutenant Governor."[50]

Mark Knouse, Scranton's administrative aide, was sitting in the room. Critchlow was shocked and looked at Knouse, whom Abraham did not know. Knouse made some expression to Critchlow to continue, and Knouse "seethed quietly."[51]

Abraham then produced a suggested edited version of Milne's first draft of a press release. Predictably, it put the blame on the state. He handed it to Milne, who read it over and then threw it away. Milne started an entirely new draft. By the time it was produced and jointly released, it was after midnight. The draft had everyone reluctantly agreeing to an inevitable release.

With the toilet war in full swing, the state's trust of the NRC was ebbing. But a fourth event of the evening helped relations to deteriorate even further. During the day, the Met Ed and NRC personnel in the control room had tried to retrace the accident to discover what had occurred. Jim Floyd, whom Gary Miller considered to be their best operator, returned from Lynchburg, Virginia, at 2:30 A.M.[52] He relieved Gary Miller, and the analysis continued. Many things were discovered during this process. For instance, Gallina first heard about the auxiliary feedwater delay on Thursday.[53]

But the issue that bothered several investigators in the control room and particularly in Bethesda concerned the core thermocouples. Most people in the control room simply did not believe them, since to believe them led one to accept that the fuel damage was massive and that the danger was growing. It was a fair point that the thermocouples were not behaving as if they were shorted, but very little in the plant was acting as it should or even malfunctioning in ways that made sense.

All uncertainty would be cleared up when the core-cooling sample was analyzed. The health physics staff succeeded in drawing it at 5:00 P.M., but Higgins and Gallina left to brief the Governor before the results were announced.[54]

When they returned, they were shocked.[55] The core sample was reading 1000 rems of radiation.[56] With that high a reading, a substantial portion of the core had to have been uncovered for a long period of time. At these levels, there were other implications of future damage that they had yet to consider.

First, Higgins had to call the state. They had just left the state with the impression that the crisis was over. More specifically, they had told Thornburgh that the core had only minor damage.

Higgins was not sure who to call. He finally called Paul Critchlow and told him that he wanted to pass along the latest information.[57] He said that the fuel damage was substantial and that the reactor would continue to be a problem for a long time. There was also the increased possibility of radiation releases.

For some reason, Critchlow understood the call to be confidential and not an official NRC report. Therefore, he interrupted the negotiations with Abraham and asked Abraham to check some confidential information that he had just learned. Abraham called Bethesda and reported back that the fuel damage was more than anticipated and that the cooling process would take longer. However, it did not mean that the low level releases would continue.[58] This was in conflict with Higgins' statement, and it did not help bolster Abraham's or Bethesda's credibility.

About 11:30 P.M., Critchlow called Governor Thornburgh to report on the negotiations with Abraham and to tell him about the Higgins exchange. For the second night, Thornburgh had a troubling thought to finish his day and to accompany him to bed. It seemed that every time that the state tried to get information from someone, the credibility of the source immediately collapsed.[59] What Thornburgh needed was one competent source at the plant that he could trust.[60]

NOTES

1. Testimony of William Wilcox, President's Commission on the Accident at Three Mile Island, April 26, 1979, p. 34.

2. Testimony of John Davis, President's Commission, May 31, 1979, p. 247.

3. Description from "Report of the Office of the Chief Counsel on Emergency Preparedness" to the President's Commission, October 1979, pp. 21-22.

4. Testimony of Douglas Costle of EPA, President's Commission, April 26, 1979, pp. 58-59.

5. Testimony of William Scranton, Pennsylvania Select Committee on Three Mile Island, May 10, 1979, p. 46.

6. Testimony of John Deutsch of DOE, President's Commission, April 27, 1979, p. 17.

7. Testimony of Richard Dubiel, President's Commission, May 31, 1979, pp. 82-84.

8. Testimony of John Deutsch, President's Commission, April 27, 1979, p. 37.

9. Deposition of Joe Deal for President's Commission, Washington, D.C., August 20, 1979, p. 76.

10. "Report of the Office of the Chief Counsel on Emergency Preparedness," p. 21.

11. Testimony of John Villforth, President's Commission, August 2, 1979, p. 264.

12. Deposition of Donald Carbone for President's Commission, Washington, D.C., August 24, 1979, pp. 49–53.

13. "Report of the Office of the Chief Counsel on Emergency Preparedness," pp. 21-22.

14. Deposition of Jack Watson for President's Commission, Washington, D.C., September 6, 1979, p. 107.

15. Deposition of Joe Deal for President's Commission, p. 12.

16. "Report of the Office of Chief Counsel on Emergency Response" to the President's Commission, October 1979, p. 9.

17. Deposition of Tom Gerusky for President's Commission, p. 34. But in testimony by Tom Gerusky, President's Commission, August 2, 1979, p. 92, he says that BRP placed the second call and asked for assistance.

18. Deposition of John McConnell for President's Commission, Washington, D.C., August 14, 1979, pp. 35–42.

19. Ibid., pp. 45–46.

20. Deposition of William Wilcox for President's Commission, Washington, D.C., August 20, 1979, pp. 26–27.

21. Testimony of William Wilcox, President's Commission, pp. 11-13.

22. Deposition of Kevin Molloy for President's Commission, Harrisburg, Pennsylvania, July 26, 1979, pp. 37–38.

23. Testimony of Herman Dieckamp, President's Commission, May 30, 1979, p. 17.

24. Testimony of Harold Denton, President's Commission, May 31, 1979, p. 302.

25. Testimony of Roger Mattson, President's Commission, June 1, 1979, pp. 81-82.

26. Deposition of William Dornsife for President's Commission, Harrisburg, Pennsylvania, July 24, 1979, p. 53.

27. Deposition of Charles Gallina for President's Commission, Bethesda, Maryland, August 2, 1979, p. 53.

28. "Report of the Office of Chief Counsel on Emergency Response," p. 26.

29. Paul Critchlow, quoted in "Report of the Office of Chief Counsel on Emergency Response," pp. 26–27.

30. Deposition of William Dornsife for President's Commission, pp. 46-53.

31. Testimony of Richard Thornburgh, President's Commission, August 21, 1979, p. 3.

32. "Report of the Office of Chief Counsel on Emergency Response," p. 26.

33. Testimony of William Scranton, President's Commission, August 2, 1979, p. 187.

34. Summation from Scranton printed in "Report of the Office of Chief Counsel on Emergency Response," p. 27.

35. Scenario from testimony of Gordon MacLeod, President's Commission, August 2, 1979, pp. 133-34.

36. Deposition of Anthony Robbins for President's Commission, Washington, D.C., July 27, 1979, pp. 35-36.

37. Testimony of John Villforth, President's Commission, p. 267.

38. Deposition of Gordon MacLeod for President's Commission, Harrisburg, Pennsylvania, July 23, 1979, pp. 22-23.

39. Ibid., pp. 31-33.

40. Content of the briefing from "Report of the Office of Chief Counsel on Emergency Response," p. 33.

41. Deposition of Richard Thornburgh for President's Commission, pp. 39-40.

42. Deposition of Charles Gallina for President's Commission, pp. 56-57.

43. Scenario from "Report of the Public's Right to Information Task Force" to the President's Commission, October 1979, pp. 155-69.

44. Deposition of George Smith for President's Commission, King of Prussia, Pennsylvania, August 17, 1979, p. 39.

45. Quoted in "Report of the Public's Right to Information Task Force," p. 162.

46. Testimony of Richard Thornburgh, President's Commission, August 21, 1979, p. 6.

47. Deposition of Richard Thornburgh for President's Commission, pp. 42-43.

48. "Report of the Public's Right to Information Task Force," p. 163.

49. Quoted in Ibid., p. 162.

50. Quoted in "Report of the Office of Chief Counsel on Emergency Response," p. 37.

51. "Report of the Public's Right to Information Task Force," p. 164.

52. Testimony of James Floyd, President's Commission, May 31, 1979, p. 189.

53. Deposition of Charles Gallina for President's Commission, p. 18.

54. Testimony of Herman Dieckamp, *Accident at the Three Mile Island, Part II*, May 24, 1979, p. 64.

55. Deposition of James Higgins for President's Commission, p. 42.

56. Testimony of Charles Gallina, President's Commission, May 31, 1979, p. 285.

57. Content of call from deposition of James Higgins for President's Commission, pp. 45-46.

58. Reported in "Report of the Office of Chief Counsel on Emergency Response," p. 39.

59. Testimony of Richard Thornburgh, President's Commission, August 21, 1979, p. 7.

60. Deposition of Richard Thornburgh for President's Commission, pp. 42-43.

Chapter 9

The Fall and Rise of Harold Denton

There are certain days in history that might best be forgotten. When the system established to maintain coordination collapses, and people do not step in to take up the slack, there are often so many poor performances that it is difficult to salvage positive lessons for the future. At Three Mile Island, that day was March 30, 1979.

But hidden within most times of complete chaos is the opportunity for natural leaders to arise. The leadership is not always permanent; the financial forces that organized to save the stock market after the panic of Black Tuesday of 1929 could sustain a rally for only two days. But at Three Mile Island, two leaders arose from the ashes of March 30 to dominate the rest of the recovery effort.

The deterioration toward Friday began inside the plant. The core coolant sample drawn back at 5:00 P.M. on Thursday and analyzed by 6:00 P.M. showed that the core was substantially damaged. The precise amount of damage was not known. The sample was drawn through a lab that had high background contamination, and then was diluted to one part in a million for analysis. There were so many uncertainties and potential measurement errors that anywhere from 20 to 60 percent of the fuel cladding could have failed.[1]

It made a sizable difference where the actual damage fit in that range, because the amount of fuel damage reflected the amount of heat generated during the accident. If the amount of heat exceeded certain levels of temperature and duration, the water around the core would separate into oxygen and hydrogen. The oxygen would then chemically combine with the zirconium on the fuel pellets, causing more heat and more releases. More important, the hydrogen would collect in the high portions of the coolant loop, where it could still block coolant flow. Theoretically, it could even explode.

There was an abundance of help available to analyze any data. However, at first, the best anyone could do was to speculate on the amount of core damage. Going into early Friday, the Met Ed employees in the control room still had the general impression that the core damage was not dangerous.[2]

The NRC had several experts who could make their own judgments. Late Thursday evening, a group of licensing personnel from the NRC arrived at the control room and began doing their own calculations.[3] The industry advisory group was still at the observation center, although they had a fraction of the personnel that they would have on site within thirty-six hours. Of course, Bethesda was busy throughout the evening. But none of these groups had the information that they needed to determine the precise amount of core damage. In the interim, beginning before midnight and continuing until the panic of midmorning, the issue before the technicians was the source of the water cooling the core.

In a reactor, there are two potential sources of core-cooling water. First, there is the series of tanks, valves, and pumps attached within a general loop called the let-down makeup system. The number of components and possible flow paths of this loop is so complex that it is difficult to envision it as one system. However, it has the characteristic that it is a set of components with a self-contained supply of water that can regulate the amount of water circulating through the core vessel. In case of leaks, which are normal, there are makeup tanks and bleed tanks and other such reservoirs that have water and storage capacity to spare.

In an emergency, there is a separate source of water for the core. Water can be drawn from the borated water storage tank that sits beside the containment building. It has several hundred thousand gallons of water, but let-down water cannot be circulated back through this tank. Once the tank runs dry, it remains dry until new borated water is shipped in and added.

Obviously, operators prefer to use the closed loop and to save the borated water supply for an emergency. But twice during this crisis, operators faced the possibility that they might not be able to use the normal water supply for the indefinite future.

Back on the first day of the crisis, flow through the normal cooling loop stopped because of steam voids. Water was still added through the makeup pumps, but the water then escaped through the pressurizer. To reach the let-down valves and complete the circle back to the makeup tanks, the water had to pass through the hot legs, which were blocked by steam voids. Since the let-down valves were drawing only steam and could add no water to the makeup tanks, Miller had to use the borated water supply to supply water to the makeup pumps.

Miller finally solved the problem on Wednesday, when he started a reactor coolant pump and forced a flow through the loop and therefore into the let-down valves. With the new flow, the makeup tanks had a continual source of let-down water to supplement the remaining flow through the normal reactor coolant loop. Borated water was still needed to force extra water through the core, but not at rates that caused the operators serious concern.

After the coolant pump was started on Wednesday evening, the core was cooled in the same mode for over twenty-four hours. Water was forced through generator A and through the core vessel by the reactor coolant pump. To supplement this flow, to assure that enough water was passing over the core to keep it covered, the makeup let-down loop was also in operation, along with minor additions from the borated water supply.

But by Thursday evening, the makeup let-down loop was in trouble. Between the let-down valves and the makeup pumps, the water system is maintained at low pressures. There is no advantage to having high pressure here, and the entire loop is dotted with pressure relief valves and waste gas vent valves to prevent pressure damage to the components. Waste gas pressure in particular can be vented, run through a compressor, and stored in the auxiliary building in a waste gas storage tank.

In normal operation, the integrity of this loop is taken for granted, and the system is used instead to relieve pressure in the primary coolant loop. But by Thursday evening, the makeup let-down system had developed two substantial difficulties.

First, it was absorbing too many gasses from the primary loop, and these gasses were accumulating in the makeup tank. If allowed to build up in high quantities, the tank would overpressurize, and its pressure relief valve would lift. Unfortunately, when the pressure relief valve opened, it would dump both the gas and the water that was needed to cool the core. The water would have to be replaced by using more of the borated water. Should the valve stick, which was not an unknown occurrence, the plant would have to return to the continual heavy use of borated water.

There was an easier way to relieve the pressure. Near the top of the tank was a waste gas vent that could be manually opened to relieve pressure without losing water. But that was the second difficulty on Thursday. The vented waste gas passed through the vent header on its way to the waste gas storage tank. The vent header had overpressurized and ruptured back on Wednesday. Every time the waste gas vents were used, radioactivity soared in the auxiliary building and the fuel-handling building.[4]

Both the pressure relief valve and the waste gas lines from the makeup tank passed through this vent header. If any pressure relief occurred, it would cause new radioactive releases to the atmosphere. With the gas pressure continually building, the relief and releases were inevitable.

The operators felt for several hours that they could minimize the releases if they relieved the pressure in small bursts. After obtaining all the needed authorizations, they began Thursday evening opening the waste gas vent valve for short periods.[5] As Floyd came on duty at midnight to relieve Mike Ross, the venting situation was explained to him.

Over the next few hours, the periodic venting continued. Radiation releases slowly built off site to about 10 millirems per hour, which was not enough for serious concern.[6] However, the slow venting was not enough to accomplish its purpose, since the pressure in the makeup tank was still building.

So far, there was no need for alarm. That would come just before 7:00 A.M., when two new major problems developed in the let-down makeup system. Instead, there was another topic that occupied the phone lines between TMI and Bethesda during the night.

Beginning late Thursday evening, Met Ed operators proposed that they should depressurize and once again try to use the residual heat pumps instead of the makeup pumps. The plant was operating at 1000 psi and would have to be dropped below 400 psi to start the pumps.[7]

There were a couple of reasons why this proposal sounded plausible. The makeup pumps were pushing only 500 gallons per minute through the core; the lower pressure pumps could push 3000 gallons per minute.[8] Much more impressively, the residual heat pumps were sturdy, and they could pump even the dirty water pulled out of the sump. The more sensitive high pressure makeup pumps required clean water, and all venting and leakage had to be replaced with the shrinking supply of borated water.[9]

Into the early hours of Friday morning, this idea was batted back and forth between TMI and Bethesda. Vic Stello in Bethesda was extremely opposed to the move. Stello was convinced on the basis of the core coolant sample that there was a hydrogen and steam void in the top of the core vessel. That void would expand during depressurization and probably uncover some or all of the core. That would heat the primary loop and fight depressurization the same way that it had on Wednesday.

In essence, there would be a race.[10] The question was whether the pressure could be dropped to 400 psi fast enough to get the residual heat pumps started before the core began to deteriorate further and prevent depressurization. If the race was lost, everyone agreed that there would be further core deterioration while the pressure was raised back to the 1000–1200 psi needed to get high pressure flow at adequate volumes.[11] The stakes were high, and the only disagreement was that Met Ed believed that the cooling loop was in better condition than did some people in Bethesda.

Two incidences finally put the depressurization proposal into a category where everyone agreed that it was to be used only as a last resort. First, they actually opened the block valve for a while on top of the pressurizer. However, the primary loop pressure did not decrease. Apparently, there was a "pneumatic volume" or pressurized void that simply expanded to fill the volume left by the escaping steam.[12]

Far more important, the pressure spike in containment back on Wednesday was finally rediscovered. It is no longer clear who actually rediscovered the pressure spike on the strip chart during the early morning hours on Friday. The NRC licensing technical staff had been reviewing all available charts since arriving several hours earlier. However, it now appears that James Floyd was told by a Met Ed employee before NRC found it.

Floyd immediately recognized the significance of what he saw. He had charts showing a simultaneous pressure spike and drop in oxygen content in the containment.[13] He felt that sequence had to be caused by a hydrogen burn and that the presence of so much hydrogen had to mean that a substantial portion of the zirconium in the core had chemically reacted with water.

There were several disagreements among the analysts that had now been cleared up. The core had obviously been uncovered for several hours back on Wednesday; the high thermocouple readings were apparently accurate; there must still be a sizable amount of hydrogen collected in the top of the core vessel. But no one could yet know how much hydrogen had collected in the core vessel or whether it threatened to push the water off the top of the core. Also, there was confusion for the next two days about whether the hydrogen could explode inside the vessel.

When Floyd saw these readings, he became concerned. If Met Ed knew of the spike on Wednesday and said nothing to NRC, there could be serious trouble for the company. He called the shift supervisor who had been in the room at 2:00 P.M. Wednesday and asked if the NRC had been told.[14] He was assured that they had been.

Floyd then showed the NRC technical staff the strip chart.[15] They agreed with his conclusions that it meant that hydrogen had burned and was still in the coolant loop. They began working on ways to measure the volume of the hydrogen "bubble," but were not able to accomplish this for almost thirty hours. They also passed the news back to Bethesda.

Throughout the evening hours, bad news had been accumulating. But in Bethesda, most of the high level administrators spent the night at home. They knew on leaving Thursday evening that substantial damage to the core had occurred on Wednesday. But the reactor had been improving all day Thursday, and there was no sense of alarm.

The news they met on arriving early Friday morning was distressing and confusing to people like Harold Denton and Harold Collins and the NRC commissioners. There was clearly an accelerated crisis. The core was in much worse shape than they thought. There was a hydrogen bubble of potentially threatening but unknown size. The plant had been venting new radiation all night.

With the hydrogen bubble in the core vessel, any move to depressurize and use residual heat pumps seemed out of the question. Yet, to bring the crisis to a focus, Bethesda then understood that there were new developments so that the plant might have to abandon its makeup system entirely and make a desperate attempt to reach residual heat pressures before repressurization and a possible meltdown started.

During the early morning, Floyd had been fighting a losing battle, attempting to keep the pressure down in the makeup tank by periodic venting to the waste gas tank. He had been concerned about overpressurzing the tank to such an extent that it would lift the makeup tank's pressure relief valve and lose the water as well as the gasses. He finally lost that battle. About 4:30 A.M., the makeup tank relief valve opened and distributed the contents of the tank between the makeup tank and a much larger reactor coolant bleed holdup tank.[16] There was now ten times the storage capacity, dropping the pressures but also the water level in the makeup tank accordingly. But to illustrate the serious nature of the waste gas pressure problem, shortly before 6:00 A.M. Floyd had an entirely new problem keeping water in the makeup tank.

While the makeup tank is the reservoir for the high pressure makeup pumps, there are additional low pressure pumps that return water to the tank to keep it filled. Pressure in the makeup tank was now so high from the trapped gasses that almost no water could be pumped into it. The makeup pumps were about to run dry because there was no water in the makeup tank.

There are two options in this case. The makeup pumps can also run off the borated water supply. In fact, some borated water was being added continually to make up for losses due to venting. But Floyd was vigorously trying to save this very low quantity of water since it was all he had left to fight unforeseen emergencies, such as another stuck valve. The other option, the one Floyd preferred, was to vent the makeup tank all the way to normal pressure.

If he did that, he could open room in the tank to pump in more water. He could also filter out the high concentrations of gasses suspended in the current water, which he understood to be like seltzer water.[17] But he also would have to vent extensively through the damaged vent header. There would be heavy radiological releases.

Floyd knew that if his plan was tried but failed, there were a couple of ways that the releases could get out of control. First, the discharged gasses would go through pipes and the bleed tank and a compressor to get to the waste gas storage tank. All of these had pressure relief valves that could open and could even stick.

Second, the waste gas storage tank had been receiving discharges for days. It was nearly filled, although there was no longer any way to find out how nearly. If it filled completely during the discharge, its pressure relief valve would open and cause tremendous releases.

Finally, there was always the possibility that the valve on the makeup tank would stick open. If it stayed open, it would eventually open all the relief valves.

With so many potential dangers, Floyd wanted to have feedback as quickly as possible on the level of the planned releases. If they were too high, he would attempt to stop them. If he lost control of the release, he wanted time to warn the local communities. Therefore, he picked up the direct phone to the emergency control station on the mainland a little before 7:00 A.M. and asked that the company helicopter hover and take readings above the auxiliary building ventilation stack.[18]

Shortly after 7:00 A.M., the helicopter was in place. Floyd opened the vent valves to start the transfer. It was not a moment of high anxiety in the control room. Floyd told one NRC official what he was doing as he opened the valves. Gallina, who was the NRC coordinator on release monitoring, did not find out for several minutes. The move just did not seem all that dangerous at the time. Only a malfunction would lead to serious releases, and precautions were being taken. Very quickly, releases from the vent stack began to register in the helicopter. They fluctuated, but peaked about 8:00 A.M. at 1200 millirems above the vent stack and began to fall. The releases had reached about one-third of Thursday's highest on-site releases. Also, the air was stagnant, and the radiation was not spreading off site. Under these conditions, Floyd and Gallina decided that the off-site consequences of the release were insignificant.[19] As a result, they assumed that the event was getting no notice off site since it deserved none.

Nevertheless, Floyd knew that the venting could still get out of control should the makeup valve fail to close on command. He finally decided that he should call PEMA to put them on guard for a possible evacuation. That decision was the snowball that grew into an avalanche.

Normally, Floyd would call the Bureau of Radiation Protection to report problems of the plant. However, he was not reporting a problem or recommending an evacuation; he was just advising PEMA to be on guard. Anyway, when he tried to call PEMA, he was unable to reach them.

At 8:34 A.M. he called Kevin Molloy, the emergency management director in Dauphin County, instead. Floyd hurriedly explained that he wanted to reach PEMA because new releases were in progress. He asked Molloy to pass the message.[20]

Molloy called PEMA and relayed the message to Carl Kuehn who was monitoring the phones. However, before PEMA could return the call, they received two simultaneous calls from TMI. One, taken by James Cassidy of PEMA, was from an unidentified caller in the control room.[21] It was reported calmly, but the caller may not have come forward since because the call was loaded with misinformation.

The caller said that the release started at 8:32 A.M., almost ninety minutes after it did. He said a site emergency had been declared, which had been true for over two days. He correctly sited the 1200 millirem reading at 600 feet. However, the caller apparently misunderstood one reading announced to the control room by Gallina of 1400 millirems at the gate and reported instead 14 millirems at the gate.

Simultaneously, Floyd reached Kuehn, who had just completed the call from Molloy. The contents of that call were a classic case of garbled communications between Floyd and Kuehn. Floyd identified himself and said that the plant had additional releases. Floyd asked if PEMA was prepared for evacuation, and Kuehn said they were.[22] By Kuehn's notes, Floyd also asked Kuehn to call the Bureau of Radiation Health (*sic*). Kuehn says that Floyd called the release "uncontrolled" and quoted the wind direction for possible evacuation.[23]

Floyd understood that he had alerted PEMA in case of trouble and that he should soon get a call from the Bureau of Radiation Protection. However, Kuehn had been stunned by Floyd's excited and almost panicky telephone voice. As he soon related to Colonel Henderson, "this guy is going ape."[24]

Within minutes, Henderson arrived at the PEMA Operations Center where the calls had been received. Comparing notes on the two calls, it appeared to those in PEMA that an evacuation might be imminent. They decided to call Floyd back to see how bad things were. Communications this time were worse than before.

Kuehn asked if Floyd was ready to evacuate. Floyd assumed Kuehn was asking if the plant could handle itself in the possible evacuation. Floyd said yes, which Kuehn understood to mean that the evacuation was now needed. With that, they hung up. The unnecessary request for evacuation had been unknowingly passed.

PEMA cannot order an evacuation, but they are responsible for passing such requests from the Bureau of Radiation Protection to the Lieutenant Governor. Since they received the call first, they passed it both ways.

Dick Lamison, the duty officer actually in charge of the phones at that point, called Margaret Reilly at 8:42 A.M. He quoted her the radiation figure, but not the rest of the message. As fate had it, she had just received a call from TMI giving the same figure. She acknowledged the report and hung up. Margaret Reilly, the one person who could have stopped this evacuation rumor quickly, missed the fact that it was spreading.

At the same time, Henderson called Scranton, and then Scranton called Thornburgh. If there was to be an evacuation, Thornburgh would have to order it. For now, Thornburgh decided to wait for more information.

Henderson next called the county directors in York, Lancaster, and Dauphin counties. He told each that the chance of an evacuation was 90 percent, so they should get into an immediate state of readiness.[25] Paul Leese in Lancaster and Leslie Jackson in York put their police and fire departments on alert.

Molloy received his message at 8:54 A.M.[26] Like the other two, he put his units on alert. But unlike the other two, Molloy had heard the panicky tone of Floyd's voice when he received the first call. He decided that an evacuation was imminent and that people needed to understand that they would have to leave in an orderly and pre-arranged pattern.

In the Harrisburg area, emergency messages about the reactor were to be broadcast over WHP radio station. As previously arranged, Molloy called the station and asked them to stand by for an important message. They agreed and told the population what they were doing.

If the rumor was not stopped soon, it would be too late to stop a massive evacuation. But the garbled communications were now carried even further.

In Bethesda, the communications were far worse than in Pennsylvania. While much of the confusion was developing in Pennsylvania, the emergency management team was not aware that venting or mobilization for an evacuation was in progress. Rather, people like Denton and Collins were piecing together any information that they heard from the site, and the site knew nothing about the activities outside.

Unfortunately, the team also heard only pieces of information from the people at the site. Denton later referred to the site as an "Einsteinian black hole" that apparently swallowed the NRC team sent on Thursday.[27]

A few minutes before 9:00 A.M., NRC engineer Lake Barrett in Bethesda picked up a rumor in the center that the waste gas tanks had filled and were venting into the atmosphere. It was a misinterpretation of the voluntary precautionary release, but Barrett quickly calculated in his head that such an involuntary release should measure about 63 curies per second if it happened and would be continuous. Curies measure the amount of radioactivity in the air, whereas rems measure dosage equivalents for people.

John Davis decided that his information was interesting and took him to see the team. The emergency management team was in a separate room separated by a glass wall, where they could see the chaos but escape it. Davis brought Barrett in, and Barrett began explaining his calculations.

With timing that would never be believed in fiction, Karl Abraham called from Harrisburg. He was the NRC press relations officer who had tried to get the Lieutenant Governor to accept the blame for the sewage release on Thursday. The state disliked him intensely now, but Critchlow in the Governor's office needed him to check the report that they had from Henderson.

Abraham called Bethesda to confirm a report from the plant that releases were occurring from the cooling towers. It was not clear how an NRC employee could believe that releases could occur "from a release point in one of the cooling towers." However, the conversation was taped.[28] He also passed the information that the release was 1200 millirems per hour. He did not know where the release was measured, but during the conversation the person receiving the call decided that it must be off site.

In the emergency management team office, Barrett was explaining his information and calculations that the waste gas storage tank was rumored to be venting and that it would release about 60 curies per second if it was open.[29] Someone in the room asked him what the off-site dosage would be in those circumstances. Barrett calculated quickly in his head and announced 1200 millirems per hour.

At this point, a phone attendant burst into the room to announce Abraham's misquoted information that the plant was reporting off-site releases of 1200 millirems per hour. The coincidence stunned the room into silence. They probably wished later that the silence had lasted.

The EPA had guidelines on safe radiation doses. By those guidelines, evacuation should be considered if the total population dose was expected to fall between 1000 and 5000 millirems.[30] It was irrelevant that these were total dosages and not hourly dosages. Because of the 1200 millirem coincidence, Bethesda assumed Barrett's rumor that this was a continual uncontrolled venting of the waste gas storage tank to be correct, and that the releases would soon drastically exceed total dosage guidelines. Everyone in the room began advocating evacuation.[31]

Denton tried to call the commissioners, who were meeting in Washington. However, there was no answer.[32] The team agreed that they needed to recommend evacuation and that they did not have much time to draw out a complicated plan.[33] Still, they had no experience with this. NRC never thought in terms of evacuation.

In gathering ideas, Denton asked Barrett what evacuation radius he would suggest. Barrett was an engineer with no training in radiation health, and he declined to answer. Denton insisted. To be safe, Barrett finally said ten miles was plenty. However, Barrett believes that the room finally settled on five miles.[34] Within a couple of minutes, Denton ordered Harold Collins to call the state with the recommendation to evacuate.

Back in Harrisburg, Colonel Henderson was sitting on an increasingly nervous emergency structure and population. Then, at 9:15 A.M., Henderson answered the PEMA telephone. It was Harold Collins from Bethesda. Now both the utility company and the NRC had bypassed the Bureau of Radiation Protection to go straight to PEMA.

Collins began by comparing notes with Henderson to see if they had the same information. Henderson's description that the release was measured 600 feet above the stack was better information than Collins had. Henderson said that they had not yet decided to evacuate. Collins said that the NRC was recommending an evacuation for ten miles in the direction of the plume. Henderson said that they would consider five miles. With that exchange, the conversation ended.[35]

Now Henderson needed to make reports up and down the chain of command again. He called Scranton, who called Thornburgh. He called Molloy to warn him that he would receive the official call to evacuate within minutes.

To Molloy, this was enough information to act. He called WHP radio and was plugged into the station so that he was on the air. He warned the population that a protective evacuation might be needed due to events at Three Mile Island. He asked schools to keep children indoors. He gave some preliminaries on what people should take with them and where they would be asked to go. He asked everyone to stay by the radio for further instructions.[36]

Within a few minutes, the Harrisburg exchanges logged 103,000 successfully placed calls. That is six times the normal load, but below the number attempted. The population got the message.

But the burden of responsibility rested with the Governor. He had recommendations passed from the plant and the NRC to evacuate. While he did not know how the recommendations started, he knew that they came from different sources. He had emergency machinery ready to move. He had a population that was nervous and wanted some kind of instructions.

Then the air raid sirens went off in the city of Harrisburg. The official story is that the siren system malfunctioned.[37] This researcher has heard rumors within Pennsylvania state government that an employee decided on his or her own to make sure that everyone knew to be on alert.

At any rate, the Governor could see the people standing in the streets of downtown Harrisburg as the siren blared for fifteen minutes. Ten blocks east, the streets out of town were jammed to a standstill with a sudden surge of panicky evacuation traffic.[38] The situation was, in a word, explosive.

With all the recommendations and the evacuation machinery almost starting itself, the pressures of the moment almost forced the Governor to act. But Thornburgh had nerves beyond mere mortals. There was something about this chain of events that struck Thornburgh as being out of place. Who was Harold Collins? Why was this new voice making such an important recommendation? He was not going to respond until he established the credibility of this Harold Collins person.

That delay bought the time that the Bureau of Radiation Protection needed. When Henderson called Scranton, another officer in PEMA called BRP. Tom Gerusky, Maggie Reilly, and William Dornsife had their own readings in progress, and the plant had earlier explained to them what was taking place. They knew that someone had fouled some communication lines, so they each called one alleged source of the evacuation request. Gerusky started trying to reach the Governor.

Dornsife called the plant and reached Charles Gallina. Gallina had been monitoring the releases for some time. Each time there was a peak, such as at 8:00 A.M. when one of the in-line pressure relief valves apparently opened, he called Region I.[39] Gallina had also called Dornsife once during the releases to keep him informed.

But shortly after Molloy's radio message, a worker came into the control room to say that the NRC was ordering an evacuation downwind. No one knew whether to believe such a preposterous statement. But he assured them that it had been officially announced on the radio and that his wife was going to the school to get the kids out of here.[40] Gallina called Region I. George Smith at Region I knew nothing, but he said that he would check into it.

Then Dornsife called, and he was "livid."[41] Dornsife demanded to know what was happening, but soon discovered that Gallina knew less than he did. From Dornsife, Gallina found out that Collins had interpreted the 1200 millirems to be continuous and off site. They both agreed that Collins had to be stopped.

But Collins had his hands full. Maggie Reilly had called him. She was furious with Collins, and she demanded to know the names of the people responsible. Collins immediately backed off, trying to protect the identity of the people on the team who had committed what Reilly called "a low blow for those turkeys."[42] By the time the conversation was over, Collins was also talking about the decision-makers as "they," not mentioning that he had participated.

Collins continued to work himself into a hole. In Reilly's call, he figured out that the state suspected that he personally had made the evacuation decision. He called Henderson back, once again bypassing the Bureau of Radiation Protection, to assure Henderson that the evacuation decision had come from the NRC and not from him personally. Collins thought that this explained things. But the obvious impact on Henderson was to impress upon him that the evacuation recommendation was now being endorsed by higher and higher authorities.[43] Henderson began passing along his communications network the information that Bethesda had reinforced the seriousness of conditions at the plant.

As far as the Bureau of Radiation Protection could see, all sense of reason out there was collapsing. They tried desperately to reach the Governor and PEMA, but all the lines were jammed. Instead, Reilly covered the phones, while Dornsife literally ran to PEMA and Gerusky started the climb up the hill to the Governor's office.[44] By whatever means possible, they intended to stop this.

In the Governor's office, Critchlow was checking the credibility of Collins. The NRC spokesman, Abraham, knew Collins but was in the dark on what had been happening. Abraham checked with Bethesda while Critchlow called Colonel Henderson. Abraham reported back that Collins had recommended evacuation, but Abraham did not think that Collins had the authority.[45] The lack of confidence put the first doubt into the Collins recommendation.

But when Critchlow called Henderson, another coincidence of timing occurred. Henderson said that Collins had a good reputation. Critchlow asked Henderson for a recommendation on the possible evacuation. Henderson said that it sounded to him as if a five mile evacuation was appropriate, but that he would like to hear from the Bureau of Radiation Protection.

At that moment, Dornsife burst into Henderson's office out of breath, but trying to get the evacuation stopped. Henderson told Critchlow about Dornsife's effort, and Critchlow told Thornburgh. Gerusky was still on the way to Thornburgh with the same message.[46]

Throughout the crisis, Thornburgh had been looking for someone outside of his circle of advisors whom he could trust. Collins was clearly not that person, so Thornburgh tried again. He called NRC Chairman Joseph Hendrie.

When Collins went to call PEMA at 9:15 A.M., Lee Gossick of the emergency management team reached NRC Commissioner Victor Gilinsky in his office in Washington. Gossick was excited, claiming that "all hell has broken loose here," and advised the commissioners to assemble around a speaker phone.[47]

By 9:30 A.M. the conversation began between the commissioners around their speaker phone in Washington and the team around its speaker phone in Bethesda. Public Affairs Director Joseph Fouchard warned them that he had heard from Karl Abraham that the Governor would expect recommendations on a possible evacuation from the NRC.[48]

The commissioners began asking Bethesda questions to form a recommendation. But Bethesda did not know the duration, exact measurements, cause, or spread of the release. Denton said that they had people on site but that getting information took hours. Denton suggested that they might have better luck if members of the team drove up and rotated shifts in the control room. In fact, Denton said,

"I would be happy to volunteer and see how things go along for a while."

The seed had been planted. But first, Hendrie needed to call the Governor. The team was not sure what to recommend. The commissioners decided to recommend an advisory to stay indoors. Hendrie called the Governor. Like everyone else, he was unable to get through.

Back at the plant, Gallina was still trying to stop all this chaos.[49] But from Region I, he found out that the decision has escalated to Hendrie, who was going to recommend an advisory to stay indoors. Gallina was trapped. He felt that the decision was wrong. A Met Ed employee ran up to him "yelling and screaming, 'What are you guys doing? Do you realize what it is going to do to us and to the industry? Do you realize what it is going to do to the people around here?'"[50] Gallina agreed, but the people at the top were now involved, and he dared not intercede.

At 10:07 A.M. Thornburgh called Hendrie. For once, a phone conversation was understood on both sides.[51] Thornburgh wanted to know if an evacuation was justified. Hendrie said that his information was terrible, but probably not. Thornburgh wanted to know the reason for the 9:15 A.M. call so that he would know how to interpret calls in the future. Hendrie said that he would find out, although Hendrie never called back about the Collins episode.

Hendrie finally advised a stay indoors advisory for five miles in a northeast direction, although he pointed out that the release was over and that most contaminated particles had already settled. Hendrie finally admitted that the NRC had probably overreacted.

Thornburgh restated the difficulties that he was having with multiple sources of communications.[52] He said that what he really needed was one high level official in the plant whom he could trust. Denton's earlier offer to Hendrie had just gained momentum.

Thornburgh had a volatile situation brewing outside, and he did not wait to draft a formal press announcement. Trusting Paul Critchlow and David Milne, he sent them out to explain to the press exactly what they knew about the crisis. They did a superb job of relating a complicated story. The only confusion was that the request to stay indoors for five miles northeast of the plant became a ten mile radius advisory.

To be sure, no one was calm yet. Dornsife and Gerusky, for instance, had just arrived at the Governor's office to try to stop the evacuation order. However, due to the incredibly well-timed caution of Thornburgh, the panic had clearly crested and was receding.

Just after 10:30 A.M., President Carter called Hendrie to see if the

NRC needed help. Hendrie said that the communications were a mess, and the President promised to get the White House busy on that. Whatever blows the White House staff reputation may have suffered in recent years, its communications network is internationally renowned.

Carter next mentioned that he thought that there should be a senior federal person on site who could speak for the government, and Carter wanted a recommendation. That was the third time that Hendrie had heard that suggestion in less than an hour. This time he responded, and he told Carter that he had Harold Denton "packing his bags."[53] Immediately after this call, Hendrie ordered Denton to go to the site.[54]

Next Carter tried to call Governor Thornburgh. To dramatize the communications problems in Carter's mind, he could not get through either. However, Carter finally made the connection at 11:15 A.M. Carter told Thornburgh of his difficulty calling Harrisburg and said that the White House wanted to run special open phone connections to TMI, the White House, the NRC, and the Governor's office. Carter also offered Jessica Mathews as a contact person at the White House.

Thornburgh thanked him, but mentioned that he would also appreciate a senior level trustworthy technician at the site.[55] Carter said that Harold Denton was on his way and would be there as Carter's "personal representative."

Thornburgh could not ask for a better commitment than that. Denton, who a few hours before had the good fortune to order Collins to make the credibility destroying phone call rather than to make it himself, was now the President's "personal representative." Denton was on his way to becoming the model for credibility and the closest thing that the TMI episode had to a hero.

NOTES

1. Testimony of Gary Miller, President's Commission on the Accident at Three Mile Island, May 31, 1979, p. 56.

2. Testimony of James Higgins before U.S. Congress, House, Committee on Insular Affairs, *Accident at the Three Mile Island Nuclear Powerplant, Part I,* 96th Cong., 1st sess., May 10, 1979, p. 113.

3. Testimony of Norman Moseley, President's Commission, May 31, 1979, p. 298.

4. Testimony of James Floyd, President's Commission, May 31, 1979, p. 167.

5. Ibid., p. 179.

6. Testimony of Tom Gerusky, President's Commission, August 2, 1979, p. 93.

7. Testimony of Joseph Hendrie, President's Commission, April 26, 1979, p. 122.

8. Testimony of James Floyd, President's Commission, May 31, 1979, pp. 226-27.

9. Ibid., p. 226.

10. Testimony of Vic Stello, President's Commission, June 1, 1979, p. 82.

11. Testimony of James Floyd, President's Commission, May 31, 1979, p. 226.

12. Testimony of Joseph Hendrie, President's Commission, April 26, 1979, p. 122.

13. Testimony of Gary Miller, President's Commission, May 31, 1979, p. 195.

14. Testimony of James Floyd, President's Commission, May 31, 1979, p. 217.

15. Mailgram from Herman Dieckamp to Richard Kennedy at NRC dated May 9, 1979, reprinted in *Accident at the Three Mile Island, Part II*, p. 191.

16. "Investigation into the March 28, 1979 Three Mile Island Accident by Office of Inspection and Enforcement," NUREG 0600 (Washington, D.C.: Nuclear Regulatory Commission, August 1979), pp. (II-3-10)-(II-3-13).

17. Testimony of James Floyd, President's Commission, May 31, 1979, p. 175.

18. Ibid., p. 181.

19. Deposition of Charles Gallina for the President's Commission, Bethesda, Maryland, August 2, 1979, p. 71.

20. Testimony of Kevin Molloy, President's Commission, August 2, 1979, p. 8.

21. Citation from PEMA log reprinted in "Report of Chief Counsel on Emergency Response" to President's Commission, October 1979, p. 43.

22. Testimony of James Floyd, President's Commission, May 31, 1979, pp. 177-78.

23. PEMA log, quoted in "Report of the Chief Counsel on Emergency Response," p. 43.

24. Testimony of Oran Henderson, President's Commission, August 2, 1979, p. 40.

25. Ibid., pp. 40-41.

26. Testimony of Kevin Molloy, President's Commission, August 2, 1979, p. 9.

27. Testimony of Harold Denton, President's Commission, May 31, 1979, p. 304.

28. Part of NRC transcript, reprinted in "Report of the Office of Chief Counsel on Emergency Response," pp. 45-46.

29. Deposition of Lake Barrett for President's Commission, Washington, D.C., July 28, 1979, pp. 54-55.

30. Testimony of Stephen Gage, President's Commission, April 26, 1979, p. 98.

31. Testimony of Lake Barrett, President's Commission, August 2, 1979, pp. 299-300.

32. Testimony of Harold Denton, President's Commission, May 31, 1979, p. 304.

33. Deposition of Harold Denton for President's Commission, Bethesda, Maryland, August 2, 1979, pp. 126-27.

34. Testimony of Lake Barrett, President's Commission, August 2, 1979, pp. 301-302.

35. NRC transcript, reprinted in "Report of the Office of Chief Counsel on Emergency Response," p. 48.

36. Testimony of Kevin Molloy, President's Commission, August 2, 1979, p. 10.

37. Testimony of Paul Doutrich, President's Commission, May 19, 1979, p. 124.

38. Personal interview with Charles Kennedy of Governor's Action Center, Harrisburg, Pennsylvania, June 18, 1979.

39. Testimony of Charles Gallina, President's Commission, May 31, 1979, pp. 267-72.

40. Deposition of Charles Gallina, p. 82.

41. Ibid., pp. 86-87.

42. NRC transcript, reprinted in "Report of the Office of Chief Counsel on Emergency Response," p. 49.

43. Testimony of Oran Henderson, President's Commission, August 2, 1979, p. 41.

44. Deposition of William Dornsife for President's Commission, Harrisburg, Pennsylvania, July 24, 1979, pp. 75-76.

45. Critchlow interview described in "Report of the Office of Chief Counsel on Emergency Response," p. 51.

46. Testimony of Tom Gerusky, President's Commission, August 2, 1979, p. 96.

47. Deposition of Victor Gilinsky for President's Commission, Washington, D.C., September 8, 1979, p. 155.

48. Taped conversation described in "Report of the Office of Chief Counsel on Emergency Response," pp. 52-53.

49. Testimony of Charles Gallina, President's Commission, May 31, 1979, p. 257.

50. Deposition of Charles Gallina, p. 93.

51. Recorded text, reprinted in part in "Report of the Office of Chief Counsel on Emergency Response," pp. 53-55, and in "Report of the Public's Right to Information Task Force" to President's Commission, October 1979, p. 189.

52. This section from testimony of Richard Thornburgh, President's Commission, August 2, 1979, p. 9.

53. Deposition of Joseph Hendrie for President's Commission, Washington, D.C., September 7, 1979, p. 230.

54. Testimony of Harold Denton, President's Commission, May 31, 1979, p. 302.

55. Testimony of Richard Thornburgh, Pennsylvania Select Committee on Three Mile Island, May 10, 1979, p. 4.

Chapter 10

Playing Hardball

For fifty hours, the control room at TMI-II operated like a big black box. Important decisions were made in there, but the outside world could not see them, find out what they were, or understand why they were made. The Bureau of Radiation Protection had better luck than most. However, both the Governor and the NRC felt virtually ignored by the plant.

Within an hour of Floyd's precautionary call on Friday morning, the roles completely reversed. People outside the plant began making the definitive judgments on the condition of the reactor and the best ways to proceed. Over the next few hours, the people in the control room learned the anger and frustration of learning only bits and pieces of what had already been decided. After 2:30 P.M., when Harold Denton arrived, many of the people who fought the early crisis picked up the unmistakable character of excess baggage.

Most of the early actors had done their part to deserve their fate, although some of the more talented people on site eventually gained the respect of the new leadership. But there was unmistakably a new leadership working on the crisis, and it spread far beyond the plant. It grew from a massive and uncoordinated mobilization that quickly arose throughout the federal government.

Three Mile Island was big international news Friday morning, and federal agencies once again scrambled for a piece of the turf in the recovery effort. Agencies that had inched into Pennsylvania over the past two days now marched in unopposed. Entirely new agencies using the flimsiest of justifications simply showed up. All previous emergency planning in Pennsylvania was quickly forgotten. On the home front in Washington, agency heads proudly proclaimed themselves to be "coordinators" of the recovery effort.

The first group to become embroiled in the search for a new TMI role was the NRC commissioners. Until Friday morning, they individually stayed at their desks or watched the recovery effort depending on their personal preferences. At 9:30 A.M, they were called together for the first time to discuss TMI in an official capacity. As the crisis escalated around them, they stayed active for the next several days.[1] In fact, Commissioner Gilinsky had a press center established at Bethesda so that NRC could disseminate its own information.[2]

The commissioners were not prepared, and they floundered at first in their new role. During the 10:07 A.M. call from Governor Thornburgh, Hendrie had promised to call him back when Hendrie discovered whether Collins' recommendation to evacuate had been justified. Hendrie and Thornburgh had agreed that the evacuation was not needed yet, but they had also agreed that the NRC would collect more information and get back to Thornburgh.

For an hour and a half, the commissioners stumbled through their sources and options, able to gather no better information than that the original recommendation was based on invalid assumptions. To at least one staff person in the room, it appeared that Hendrie did not want to make a decision, but preferred to stall and wait for more information.[3]

Unknown to the commissioners, events were developing inside the NRC staff and at John Herbein's 11:00 A.M. news conference at the plant that threatened to leave the commissioners behind. However, having decided very little, Hendrie finally called Thornburgh at 11:40 A.M. as he had promised.

For a while, the conversation was uneventful.[4] Hendrie apologized for the earlier inaccurate recommendation. Hendrie said that the NRC still did not know what was happening at the plant. That led to a discussion of the communications systems being installed and of Harold Denton's new role.

Then Thornburgh mentioned that Pennsylvania's Secretary of Health Gordon MacLeod had earlier recommended evacuating small children and pregnant women.[5] The subject had also arisen during the commissioners' discussions, and Hendrie agreed that it was probably a good idea, since they knew so little about the long-term stability of the reactor. They debated possible distances of evacuation for a while and settled on a distance of five miles.

After the phone conversation was completed, the advisors in the Governor's office began working through the details of a public advisory on evacuation. Although the danger of genetic damage lessens after infancy, they agreed that including all preschool children would be safer and would seem more logical to concerned parents.[6] Since parents probably would not evacuate the area with preschool children and leave older children in school, the Governor decided to close all schools within a five mile radius of the plant. There was a delay while the staff checked to verify that the Governor had the authority to close schools. Then, about 12:30 P.M., the wording of the advisory was completed and released.

Back at the NRC, the commissioners continued to discuss the partial evacuation advisory and other possible moves until 12:40 P.M. Then they were hit with devastating news. On Thursday, NRC's Vic Stello had decided that there must be a void in the core vessel. When Roger Mattson arrived at Bethesda on Friday morning, he combined this suspicion and the news of the pressure spike and felt certain immediately that there was a sizable hydrogen "bubble" in the top of the core vessel.[7] This same conclusion was also reached in the control room, where they did not believe the bubble to be overly dangerous. Still, the NRC on site was attempting to measure the size of the bubble.

The NRC first issued a news release mentioning the bubble at 9:50 A.M. and described it as being "of interest."[8] But there was another rumor circulating through the staff at Bethesda that made it much more than just "of interest." For several hours on Thursday evening, Mike Ross and then Jim Floyd of Met Ed talked about depressurizing to use the residual heat pumps.

By Friday morning, Vic Stello in Bethesda mistakenly understood that they had been saying that the makeup system was about to open for continuous releases at any moment and that the operators would be forced to attempt to reach 400 psi or lose all core cooling.[9] Met Ed never said any such thing since it was not true. However, Floyd had verbally considered the possibility that such an accident could occur through a stuck pressure relief valve, and that bit of speculation may have been relayed back as a prediction by someone at the plant to Bethesda.

Even if that fear was not well founded, and it was not, Stello and Mattson correctly understood that the hydrogen bubble presented another potential danger. If a portion of it separated and flowed through the loop all the way to the reactor coolant pump, that pump would stall because of the void. These pumps had already stalled on Wednesday due to steam, and only one could be restarted. Hydrogen would almost certainly knock it back out of service. If this happened, only the makeup pumps would be available to cool the core, and the makeup system was known to be having problems. Therefore, the complete loss of core cooling was not an insignificant possibility.

By 12:40 P.M. Mattson had worked himself into a state of considerable anxiety. He called Hendrie and spoke to the commissioners over the speaker phone.

> They can't get rid of the bubble. They have tried cycling and pressurizing and depressurizing. They tried natural convection a couple of days ago, they have steamed out the pressurizer, they have liquided-out the pressurizer. The bubble stays.[10]

He went on to tell them that any attempt to depressurize now would run a substantial risk of a meltdown. Mattson was strongly in favor of a forced evacuation of the population to a ten mile radius. He felt that any attempts to move the bubble ran too great a risk if the population was still there.

After talking to Mattson, the commissioners took a while to digest what he had said. They then discussed whether Thornburgh should be told about the bubble. But the discussion and inaction mode continued among the NRC commissioners, and they were clearly not making timely decisions.

Finally, Hendrie had to leave for a 1:30 P.M. meeting to brief the staff at the White House. When he left, the commission still had not decided whether to tell the Governor about the bubble. This time, the federal bureaucracy would not wait. Before the NRC got around to calling Governor Thornburgh, the White House staff did.

Throughout much of the rest of the federal government, agencies that had been thwarted earlier and agencies that never considered a role earlier began assigning themselves leadership positions. The agency with the longest seniority on the site was the Defense Civil Preparedness Agency (DCPA). Since they funded about half of PEMA's budget, they had sent a representative to PEMA on Wednesday morning without invitation. For the most part, the representative stayed out of the way. However, DCPA pointedly offered additional help on Thursday, and Colonel Henderson pointedly declined.[11]

On Friday, DCPA called again. By this time, Henderson knew that the federal government was discussing evacuations beyond the five mile radius of PEMA's plans. Since the counties were scrambling to adjust, Henderson agreed to assign two DCPA representatives to each of the four counties surrounding Three Mile Island.

The Federal Disaster Assistance Administration (FDAA) Administrator William Wilcox also tried to intercede on the first day. He pressured Regional Director Robert Adamcik to go to the site. Adamcik had reservations about barging in uninvited and called Colonel Henderson. On being told that Colonel Henderson did not need help, Adamcik stayed in Philadelphia.

On Thursday, Wilcox called Adamcik again. This time, there was an "indication just short of insistence. . ." that Adamcik or a representative go to the scene.[12] Still Adamcik stayed in Philadelphia.

On Friday, Wilcox called Adamcik again. This time he "ordered" Adamcik or a representative to go to the site. Wilcox also specifically ordered Adamcik not to ask Colonel Henderson's permission.[13] Adamcik did call Henderson to inform him that FDAA was on the way. Henderson politely responded that FDAA was always welcome.[14]

But the most obtrusive and the most resisted federal emergency management agency was the Federal Preparedness Agency. FPA Acting Regional Director Thomas Hardy called PEMA Wednesday morning to offer assistance. When he was told that none was needed, he did not press the point. But as described in Chapter 8, the FPA had been trying for years to muscle in on FDAA's role as leader of the emergency response effort. When the national office of FPA decided to move again on Friday, they instructed Hardy to convene a meeting of all federal agencies to coordinate a federal response.[15]

Hardy called Colonel Henderson to inform him of the planned meeting. But Henderson raised considerable objections with FPA and with the Governor. As Hardy understood Henderson's position, PEMA was getting swamped with federal observers milling around the emergency center. It was bad enough trying to get work done with all those people causing congestion and constantly interrupting with questions. But when one of those observers started trying to give orders to the others, emergency response was likely to break down until everyone defined and protected their political turf.

The FPA meeting never actually took place. Before they could get everyone together, the White House intervened to cancel the meeting and to write its own ground rules.

DCPA, FDAA, and FPA were just the emergency response agencies that were jockeying for rights within that political arena. But there were other agencies and at least one cabinet department that also claimed jurisdiction and that played politics equally vigorously but with more resources.

Two of the largest agencies joined resources on Friday to try to reorient the recovery effort toward an expertise in public health.

Both agencies had internal offices that gave them a claim to the area, and together they played a much larger role than their related offices would seem to have deserved.

The largest was the federal Department of Health, Education, and Welfare (HEW) under politically adroit Secretary Joseph Califano. For the first two days, Califano's involvement was restricted to his awareness that HEW's Center for Disease Control had been notified by the state of Pennsylvania.

But around 8:30 A.M. Friday, Secretary Califano was contacted by a U.S. senator who wanted to know what role HEW was playing.[16] Califano did not know, but he was determined to find out. He had his executive secretary, Rick Cotton, contact various health officers in HEW and Administrator Douglas Costle at EPA. He learned that the Food and Drug Administration was sampling food in Harrisburg and that the Center for Disease Control was still on call. Other than that, HEW appeared to have no response.

Califano was concerned that the radiological monitoring seemed dominated by the "pro-nuclear cabal" of DOE, NRC, and the utility.[17] At the least, he wanted an input to allow HEW to gather information for future health studies. At the most, he was not sure what role he wanted.

During the afternoon, Califano and Costle agreed that HEW should sponsor an interagency meeting at Califano's office at 5:00 P.M. to assemble a public health response plan. Before that meeting, Califano held a meeting of his health experts, which he sometimes called his "health cabinet."[18] These included the directors of the National Institutes of Health, the Center for Disease Control, the Food and Drug Administration, and the National Institutes of Occupational Safety and Health. At the meeting these agencies and Califano divided the response functions within HEW.

As the afternoon progressed, Califano and Costle agreed to expand their interagency contacts by extending additional invitations to the 5:00 P.M. meeting. Two NRC commissioners finally attended, as did Jessica Tuchman Mathews from the White House.

EPA meanwhile was involved in its own preparations. Its most relevant subunits were its Office of Radiation Programs and its Office of Research and Development. Between the two, EPA could provide about thirty monitors, an instrumented aircraft, and over twenty staff members. Some of this was in the Office of Radiation Programs, which was on alert as an IRAP signatory. But Costle decided to send the other equipment under the leadership of Dr. Stephen Gage. Gage did not know of IRAP and directed his team's activities on arriving on an *ad hoc* basis.[19]

At the 5:00 P.M. HEW meeting, there were perhaps thirty to forty people, although it was difficult to count with so many and with people continually arriving and leaving.[20] During the meeting there were three major topics. The health people pressed NRC to tell them how much notice would be available in case of an evacuation. Commissioner Gilinsky finally guessed about six hours.[21] EPA and HEW agreed that they would send representatives to the incident response action center at NRC in Bethesda to have health people at the source of the information.[22] Also, the health participants agreed to try to find large supplies of potassium iodide, which might be a useful medicine during an emergency evacuation.[23] Califano also mentioned that he had some personal recommendations for the White House.

After the scheduled meeting, several HEW people stayed behind to draft a statement for the White House. Among other things, the statement asked the White House to draft a memorandum for the public incorporating the views of the health officials.[24] But at the White House, other things were brewing.

The early role of the White House was surprisingly informal. On Wednesday, NRC Commissioner Gilinsky had decided that someone at the White House ought to be informed. It was not clear whom he should tell. However, he happened to be friends with Jessica Tuchman Mathews of the National Security Council staff, who had a Ph.D. in biophysics and who had discussed nuclear issues with him before.

Mathews prepared a brief memorandum of what she had been told and took it to her boss, Zbigniew Brzezinski. He asked her if the event was major, and she said that it was too soon to know.[25] He asked her if the President should be told, and she said yes. He directed her to keep informed, and then he took the memorandum to President Carter.

There was nothing more of presidential importance to report until Friday morning, when Commissioner Gilinsky called again. He told Mathews that a serious, uncontrolled release had occurred at the plant.[26] She then asked Brzezinski to tell the President while she tried to collect more information.

When Brzezinski passed the information to the President, there was still too little known for certain. Therefore, at 10:30 A.M., Carter called NRC Chairman Hendrie. That call has already been discussed. During it, Carter agreed to use White House facilities to set up a communications network. He also asked if communications would be improved by sending a senior representative to the site, and Hendrie suggested Denton.

There are certain amenities that go with White House support, not the least of which was the communications network. However, Carter also provided a helicopter. At this point, the NRC was one of the few federal response teams still using cars to reach the site.

During the meeting and Hendrie's conversation, the President was painted a picture of communications confusion.[27] This was reinforced when he tried to call Governor Thornburgh and could not get through. With all this confusion, Carter asked his staff to check into the feasibility of NRC taking control of the plant. He also wanted a meeting of federal agencies at the White House to outline their roles and to maximize their availability for the response effort.

Jessica Mathews recorded his requests and set to work implementing them after the President adjourned the meeting. The meeting of relevant agencies was set for 1:30 P.M., and Brzezinski's staff began calling all the related agencies that crossed their minds. They were not aware of the IRAP list either.[28]

So far, information about Three Mile Island had passed through the National Security Council because Mathews happened to know Gilinsky. But the federal-state coordination that was required would more logically fit the traditional political role of presidential advisor Jack Watson. While Mathews stayed with the crisis because of her nuclear knowledge, the President directed that she should brief Jack Watson and his deputy, Eugene Eidenberg. The briefing took place at 11:30 A.M.

At 1:15 P.M. Mathews prepared another memorandum for Watson to give to the President, answering several of his earlier questions. It said that Met Ed agreed not to take any actions without NRC approval, so seizing the plant served no purpose.[29] It said that the hydrogen bubble threatened to increase fuel damage. Specifically, more water was needed to cool the core. More volume could not be added without depressurizing, during which time more of the core would be uncovered.

Mathews stated that she did not know whether Thornburgh had issued the partial evacuation advisory yet. It never occurred to her that Thornburgh might still not have been told about the bubble.

Carter received the memorandum before the 1:30 P.M. White House meeting began. Carter did not attend that briefing. However, Carter had training and some work experience with navy reactors. Based on that training, Mathews' memorandum spelled out enough information for Carter to understand the situation as well as almost anyone. Fortunately, the situation then understood by the experts was significantly worse than what we now know about the condition of the plant at that time.

The 1:30 P.M. meeting was attended by Watson, Eidenberg, and Mathews from the White House. The meeting also drew representatives of the Defense Department, the Joint Chiefs of Staff, the Department of Energy, the Food and Drug Administration, FDAA, and DCPA. Some obvious agencies were omitted because Brzezinski's staff had to create a list of invitees without benefit of the IRAP.

Hendrie began the meeting with a briefing on the reactor's condition. He mentioned a "few percentage" chance of massive releases that would require a twenty mile evacuation downwind with six to twelve hours notice.[30]

The meeting then turned to organizing the chain of command in the federal effort. Watson announced that he was the White House coordinator and that the National Security Council would move out of this area that Brzezinski had never really wanted. It was agreed that Denton would be the sole source of information on the conditions at the plant. Finally, Watson designated FDAA as the coordinator of evacuation planning. FPA was not at the meeting, but they were quickly informed. Something of a federal triumvirate of contact personnel was announced consisting of Watson at the White House, Denton at the plant, and FDAA's Adamcik, who had finally gone to Harrisburg.

Other agencies still had important roles. DCPA kept their assistants on emergency planning in the affected Pennsylvania counties. But it was now clear who did the coordinating and interpreted the data. In a government that lives on politics, it took some agencies a day or two to learn that Watson was serious. However, Watson had no reservations about using White House pressure to make that point clear.

When Jessica Tuchman Mathews returned to her job, she called Thornburgh's assistant, Jay Waldman, to tell him that Jack Watson was now the White House contact person. When she began briefing him, she found out just how little the state had been told.[31]

She told Waldman that the problems in the TMI reactor were unprecedented. She said that the worst case scenario was a meltdown in four to six hours. This was nothing like what Waldman had heard previously.

Shortly after Harold Denton reached the control room, Mathews called the state to relay what information she had received from Denton. This time, she reached Thornburgh directly. She assured him that the venting that morning had been intentional and under control. However, the more recent news was far more serious. There was a hydrogen bubble threatening to uncover part of the core and to cause the condition of the plant to deteriorate. The extent of the danger was uncertain, since the people at the plant still were not sure what they were seeing in their readings.

Mathews' call to Thornburgh was quickly followed by one from Watson to Thornburgh. Watson informed Thornburgh of all the contact people. He also assured Thornburgh that the contact people were for his convenience and were not intended to restrict his sources.

Back at the NRC, Hendrie returned from the meeting to find that no evacuation update had been given to the state. However, the news was deteriorating by the minute. Roger Mattson called the commissioners in Hendrie's absence to report that they may have found a way to fight the bubble, but that it "is a failure mode that has never been studied. It is just unbelievable."[32]

Mattson insisted that evacuation was essential. However, he did not know what that entailed. Collins, who had no familiarity with the area, felt that the small towns within any ten mile plume could be evacuated in under an hour. Harrisburg might take two hours.[33] The NRC had a considerable amount to learn about evacuations.

Now that Hendrie was back from the meeting at the White House, he began briefing the other commissioners on what had been said. Specifically, he reported that decisions about possible evacuation recommendations were to be restricted to Harold Denton. Commissioner Aherne told Hendrie of Mattson's report and suggested that the NRC might want to tell Thornburgh that evacuation was a possibility.

Later in the crisis, Hendrie was to regret that he had not listened more carefully to Watson on Friday. However, he had just returned from the White House, and the instructions were still fresh in his memory. Denton had now been at the plant only about thirty minutes. Hendrie was going to give him a chance without outside interference.

NOTES

1. Testimony of Lake Barrett to President's Commission on the Accident at Three Mile Island, August 2, 1979, pp. 313–14.

2. Deposition of Victor Gilinsky for President's Commission, Washington, D.C., September 8, 1979, pp. 118–20.

3. Deposition of Thomas Gibbon for President's Commission, Washington, D.C., August 20, 1979, pp. 37–38.

4. Contents described in deposition of Tom Gerusky for President's Commission, Harrisburg, Pennsylvania, July 24, 1979, pp. 62–67, and in testimony of Richard Thornburgh, President's Commission, August 21, 1979, pp. 11–12.

5. Testimony of Gordon MacLeod, President's Commission, August 2, 1979, p. 14.

6. Deposition of Tom Gerusky, pp. 67–69. Testimony of Richard Thornburgh, President's Commission, August 21, 1979, pp. 12–13.

7. Testimony of Roger Mattson before U.S. Congress, House, Committee on Interior and Insular Affairs, *Accident at the Three Mile Island Nuclear Powerplant, Part I*, 96th Cong., 1st sess., May 9, 1979, p. 7.

8. NRC Preliminary Notification, March 30, 1979, 9:50 A.M.

9. Testimony of Vic Stello, President's Commission, June 1, 1979, p. 84.

10. With transcribing errors corrected, from NRC Hearing Transcript, March 30, 1979, p. 62.

11. Deposition of Oran Henderson for President's Commission, Harrisburg, Pennsylvania, July 30, 1979, p. 50. Deposition of John McConnell for President's Commission, Washington, D.C., August 14, 1979, p. 46.

12. Deposition of Robert Adamcik for President's Commission, Washington, D.C., August 8, 1979, p. 40.

13. Deposition of William Wilcox for President's Commission, Washington, D.C., August 20, 1979, p. 31; and deposition of Robert Adamcik, pp. 42–44.

14. Deposition of Robert Adamcik, pp. 42–44.

15. Deposition of Thomas Hardy for President's Commission, Washington, D.C., August 8, 1979, pp. 40–43.

16. Early Califano activities from deposition of Richard Cotton for President's Commission, Washington, D.C., August 16, 1979, pp. 7–12.

17. That expression to describe the common impression from deposition of Stephen Gage for President's Commission, Washington, D.C., August 13, 1979, p. 66.

18. Testimony of John Villforth, President's Commission, August 2, 1979, pp. 233–34.

19. Deposition of Stephen Gage, pp. 29–34.

20. Deposition of Jessica Tuchman Mathews for President's Commission, Washington, D.C., August 23, 1979, pp. 76–78.

21. Deposition of Richard Cotton, pp. 23–24.

22. Testimony of John Villforth, President's Commission, August 2, 1979, p. 246.

23. Deposition of Richard Cotton, p. 24.

24. Ibid., pp. 28–29.

25. Deposition of Jessica Tuchman Mathews, pp. 11–12.

26. Ibid., pp. 21–23.

27. Content of meeting from Brzezinski as related in ibid., pp. 26–29.

28. Deposition of Jack Watson for President's Commission, Washington, D.C., September 6, 1979, p. 107.

29. Memorandum reprinted in "Report of the Office of Chief Counsel on Emergency Response" to President's Commission, October 1979, p. 67.

30. Notes from meeting restructured in ibid., p. 68.

31. Conversation recounted in ibid., pp. 76–77.

32. NRC meeting transcript, March 30, 1979, p. 77.

33. Ibid., p. 96.

Chapter 11

Blowing the Bubble

At 2:30 P.M. Friday, an army helicopter set down on the roped off field behind the parking lot at the observation center at Three Mile Island. Harold Denton and several of his assistants stepped out. The observation center is not designed to conduct business and has only one small office that is not well isolated. Nevertheless, Met Ed employees took Denton to this location to meet with John Herbein so that Denton could begin collecting information on the condition of the reactor. It was difficult to talk in the room with the milling crowd, the glass doors, and the "rather chaotic situation."[1]

Next door to the observation center is the private residence of Dewey Schneider, a nonnuclear district manager for Met Ed. In fact, the observation center parking lot and entrances literally surround his house. The Schneiders were still at home, and they quickly became part of the recovery effort. Some limited meetings were conducted in their living room; their laundry room was the only suitable photographic darkroom off the island; the garage became the company's cafeteria for a couple of days, with Mrs. Schneider doing much of the cooking.[2] When that was not enough, feeding the mass of people resembled a continuing chain of fraternity pranks. In at least one incident, an employee was sent to a fast food chain to buy 400 pieces of fried chicken on credit. Within a day, the situation became so hectic that the Schneiders moved to a motel in Hershey and left their house in the company's care.

To try to get out of the chaos of the observation center, Herbein began walking Harold Denton toward "Dewey's house." On the way, they met Met Ed President Walter Creitz, who was standing on the observation center lawn. They stopped and talked for a while on the lawn since "(i)t was the only place to meet."[3]

During the morning, the hydrogen bubble had become a hot issue in the news. At 11:00 A.M. John Herbein held a press conference during which his opening statement did not volunteer any information on the bubble. When asked why core cooling was taking so long, Herbein mentioned the bubble, but said that the core was covered and was being slowly cooled.[4] In the interim, a press briefing in Bethesda by two staff members confirmed the bubble and presented the worst case scenario that it could lead to a meltdown.[5]

The disagreement of the two stories was based on the difference of perspective that the utility was concentrating on the plant being stable, even if they had no immediate possibility of depressurizing. NRC noted that while the plant was stable, either the inevitable planned or an accidental depressurization would lead to additional fuel melt. Creitz showed Denton a proposed joint GPU-NRC press release that minimized the potential of a meltdown. However, Denton made it clear to Creitz immediately that he was going to speak in the name of the NRC only.[6] GPU finally issued the release in its own name.

Denton finally made it to Schneider's house, where he met GPU Vice-President Robert Arnold. They briefly discussed the situation. They were interrupted when the kitchen phone rang, and it was President Carter wanting to talk to Harold Denton. Carter wanted to impress on Denton that the crisis had been marked by too large a concern for falsely reassuring the public, which had excited them even more. Carter wanted Denton to be completely honest in his press statements.[7] Denton agreed and said that he would be getting back to Carter with some names of industry experts whom he would like to have flown in to assist.

Denton next went to the NRC trailer to discuss with his aides whether an NRC seizure of the plant was in order. They discussed it briefly, but had no evidence that such a move was needed.[8] So far, Met Ed seemed cooperative. They therefore dismissed the idea and compared what notes they had on the plant.

Denton stepped out of the trailer to the considerable clamor of the press. He told them that he had just arrived and would have to get back to them. Then, a little before 3:00 P.M., he got into a Met Ed car and rode over to the island.

In the Unit II control room there was an abundance of advice available for Denton. Some NRC people had been there for one or two days. Some of the staff that Denton brought with him had also preceded him onto the island. From the time that Denton arrived, he had already decided that the Met Ed workers did not have the technical expertise to be of much help.[9]

Denton needed first to assess the situation quickly so that he could call Bethesda and the Governor. He asked about the cause of the deliberate venting that morning. He questioned the Met Ed people about a variety of subjects and satisfied himself that they now appreciated the potential dangers and had some preliminary plans to deal with them.[10] The reactor was being cooled by one makeup pump and one reactor coolant pump. The pressure was 1000 psi and the coolant was running under 300° F. However, one core thermocouple was still at saturation temperatures, so steam was still forming. The coolant sample and the pressure spike also gave unavoidable evidence of a bubble. Depressurization at this point was out of the question.

At 3:16 P.M. Denton called Hendrie and relayed the technical information. Denton advised against an evacuation until there was more evidence that anything was out of control. The two agreed that first Hendrie and then Denton should call the Governor.

Hendrie called Thornburgh around 3:40 P.M. and gave a technical summary of conditions to the group gathered around the Governor's speaker phone. The Governor's group was quickly becoming adept at technical terms. However, when Tom Gerusky of the Bureau of Radiation Protection heard of the bubble, it stunned him. He asked a question that put him several hours ahead of the NRC: "What are the potentials for an explosion that would rupture the core? Rupture the vessel?"[11] Hendrie replied that there was not enough oxygen in the core to spontaneously combine, and that reassurance is now believed to have been true. However, Hendrie would later realize that the answer was misleading, since oxygen was still being formed.

Denton's phone call to the Governor was placed at 4:05 P.M. Denton repeated some of the major technical points. The main thing that Denton established was that there appeared to be no difficulty in maintaining this mode of cooling for days while they decided what to do about the bubble. Thornburgh asked if Denton could come to Harrisburg later for a meeting away from the site. They agreed that 7:00 P.M. would give Denton enough time to get things settled at the plant.

While they figured out what to do, Denton wanted the NRC personnel to interview the Met Ed operators at some length. Any additional information they got from this process might be helpful. However, the operators were still busy at the controls, and Met Ed objected to the interruptions. They finally agreed that the operators would be interviewed as they finished their shifts.[12]

Harold Denton also wanted hydrogen recombiners hooked up and started in the containment building. These are machines that are able to absorb chemically any hydrogen in the air to prevent it from building to potentially explosive levels. On Friday there was no thought that the hydrogen left in containment was at potentially explosive levels. However, there had been a hydrogen explosion on Wednesday, and Denton wanted to prevent a hydrogen buildup that could cause that to happen again.

Hydrogen recombiners work very slowly, and on Friday, the one unit available could be expected to do no better than keep the hydrogen at current levels.[13] However, Met Ed could not get it started. NRC regulations required that one recombiner be stored on the site, and it was. The regulations also state that the recombiner need not be kept connected to a power supply, but that the utility must be able to hook it up quickly. That would also have been possible except that no one could enter containment to make the necessary connections.

Over the next few days, the hydrogen recombiner issue grew, as Harold Denton became convinced that the company was not making a serious effort to get the recombiners started.[14] During this period, a second unit was flown in, as well as roughly one million pounds of lead bricks to serve as a shield for workers trying to make the necessary connections. But for now, the single recombiner sat idle, and the hydrogen levels in the containment building climbed slowly.

Back in Washington, White House Press Secretary Jody Powell was attempting to form the same triumvirate for press releases that the 1:30 P.M. meeting had established for governmental information. Specifically, he wanted technical press releases to flow through the NRC, state activities and evacuation planning to come from the Governor's office, and federal activities to be described by Jody Powell's office.[15]

Through much of the afternoon, explicit instructions on this point were telephoned to the relevant agencies. Powell's office also scheduled its first press conference for 5:15 P.M. However, indications that such coordination would be difficult started about 3:30 P.M., as two staff members from Bethesda addressed a press gathering on technical issues. When asked about the possible worst case scenario, they used charts and meticulously described the process by which an accidental depressurization could lead to a meltdown.[16]

The press jumped on the word "meltdown" immediately, and UPI released the story nationally. Jody Powell then had to find the words at 5:15 P.M. to point out that the experts believed that the chance of a meltdown was remote. In addition, a standard press release put out by the lower level staff at NRC at 4:15 P.M. on the same subject was not as carefully worded and gave Jody Powell additional problems during his press conference.

Finally, after the 5:15 P.M. press conference, Powell called the NRC commissioners. In light of the problems related to the two NRC press statements, he asked the NRC to be a little more careful in what they said to the press. In addition, he asked the commissioners to cancel the two television appearances that they had scheduled for that evening.[17]

Back in Pennsylvania, Denton also received heat for the NRC press releases. As he related back to the commissioners, the "Governor's office calls every few minutes."[18] This delayed him considerably in getting his own decision-making structure established in the plant. By the time he got to the Governor's office for the planned 7:00 P.M. briefing, it was 8:30 P.M.

At the Governor's briefing, Denton reported that he felt that there was no immediate danger from the bubble and that adequate warning would be available if it got out of control. However, should something go wrong, a twenty mile evacuation might be needed. Colonel Henderson, who was at the briefing, had quite a task ahead of him updating his five mile radius evacuation plans.

After the briefing, Thornburgh and Denton appeared together in a news conference. After Thornburgh's opening statement, the very first question was on a technical issue. Thornburgh turned to Denton, who answered the next several questions, and a new reputation of public trust was started.

In reading the transcripts of Denton's statements, it is difficult to identify what it was that created this aura. Like previous spokesmen, time often proved him to be incorrect. The reactor was sometimes in better shape and sometimes in worse shape than he believed. He was often vague. But his casual speech and style of delivery made him appear sincere.

A lack of accuracy in the control room was perhaps not as threatening to the decision-makers and the watching world as a lack of honesty or competence. A sincere, competent individual could at least admit that problems existed and could work through them if there were solutions. Beginning Friday evening, Denton seems to have convinced the state and the press that he was such an individual.

Before Denton returned to the plant, new trouble was brewing in Bethesda. During the early evening, it occurred to Chairman Hendrie—as it had occurred to Tom Gerusky in the Governor's office—

that no one was monitoring the possibility of an oxygen buildup that might lead to a hydrogen explosion in the core vessel. It was a theoretical danger that might prove to be unfounded, but Hendrie had the distinct impression that no one was checking.

At 9:30 P.M. Hendrie called Roger Mattson at the center to see if anyone was working on that topic.[19] The topic caught Mattson by surprise. However, since oxygen was being continually released into the core, and a concentration as small as 4 percent was theoretically flammable, Mattson became intrigued. He agreed to check into it.

About 2:00 A.M. Saturday Hendrie called the plant. Vic Stello received the call, since he had relieved Denton for the evening.[20] Hendrie knew that several nationally respected nuclear physicists had already gathered at the plant and that they might be able to answer his question on the oxygen buildup. Stello agreed to have them work on it under the supervision of NRC's Matthew Taylor.

There was one other major development during the early hours of Saturday morning. The NRC technical staff finally arrived at figures on the measurement of the bubble. They decided that it must be 1000–1500 cubic feet and pressurized to about 1000 psi. Because of continued instability in parts of the core, it was growing. However, the core was still covered by several feet of water, and the bubble was growing at less than 50 cubic feet per day.[21] At this rate, there were days left to decide what to do. But all that was predicated on the assumption that the core vessel would not explode.

As the early morning hours of Saturday progressed, there was a growing split between the people in the control room and the people in Bethesda on whether the bubble had the possibility of exploding. The split mushroomed as some of the primary actors, most notably Denton at TMI and Mattson at Bethesda, returned to work the next morning.

Mattson returned to his post about 9:00 A.M. Saturday to find that there was still no answer on even the fundamental question of whether oxygen was still being generated. Hendrie had asked Darrell Eisenhut at Bethesda and Matthew Taylor to work on this, but Hendrie had no answers yet. Mattson added two separate NRC groups to the investigation, with the explicit permission to use outside consultants as appropriate.[22]

Hendrie, who had spent the night working on this problem, called Mattson at 10:30 A.M. to spur him along. Both had three concerns. First, should the oxygen build to explosive levels, it could rupture the core vessel with some internal projectile shaken loose by the explosion.[23] Should this not happen, it could jar the core structure loose, possibly into a molten mass. Finally, there was enough hydrogen in the core that it could even partially ignite, leaving enough for a secondary explosion to breach containment.

The possible consequences were obviously very severe, so Mattson's teams continued to work. Meanwhile, the commissioners debated telling the Governor, with no apparent consideration in the meeting transcript that Watson at the White House had ordered them to forward any such communications through Denton. Finally, they decided to call Denton, and he promised to relay their concern to the Governor.

For the moment, Bethesda waited to receive their staff reports. Finally, about 1:00 P.M. and 2:00 P.M. Saturday afternoon, they received preliminary reports from each of Mattson's studies. One said that the bubble contained 2 to 3 percent oxygen and would reach an explosive 5 percent within four to five days. The explosion could generate a devastating 20,000 psi spike in the primary loop, which would blow the entire cooling system apart. The other report, from Westinghouse consultants, also said that oxygen was being added to the coolant. Westinghouse did not believe that an explosion could happen at this time, but they had not yet finished enough calculations to rule it out.[24]

The reports had several warnings written in, and the Westinghouse report was both incomplete and negative. But neither report ruled out what the commissioners and Mattson had feared. Mattson says that he was not aware of two other reports that James Taylor of Babcock and Wilcox was circulating around the center at Bethesda. One said that no excess oxygen was being generated at all since free hydrogen in the coolant was absorbing it as fast as it was generated. The other said that the dampening effects of the steam in any explosion would lessen the pressure spike to 3000 or 4000 psi, which the loop could withstand.[25] Mattson was aware of only this second finding, which had the difficulty that it was from Babcock and Wilcox. They obviously had a vested interest in the integrity of their steam system.

Based on what they agreed was the thrust of Mattson's findings, Hendrie called a 2:30 P.M. press conference. It was a decision that bordered on insubordination, given earlier White House directives and a decision that he regretted deeply later.[26]

During the press conference, Hendrie discussed the several methods being tried to get the bubble out of the core vessel. He said that a precautionary evacuation of ten to twenty miles might be advisable while the bubble was being manipulated. In response to a question, he said that the bubble would become "flammable," intentionally avoiding the more frightening word "explosive," only if the oxygen built up over time, as it seemed to be doing. In addition, he said, trying to manipulate the bubble could increase the danger of explosion. But when reporters began to ask about conflicts between his story and the one presented by Herbein less than an hour earlier, the muddled response made it clear to many that Hendrie knew less about current developments than those at the plant and that his information had the potential of confusing the public even further.[27]

Back at the plant, there was some disagreement between Denton and Met Ed on just how serious the explosion danger was. However, neither side was subject to the wild speculations from incomplete data that seemed to be epidemic at Bethesda.

Denton had left the reactor Friday night, much more concerned with the possibility that the hydrogen bubble might block the coolant loop than with the worries that arose Saturday.[28] But on arriving Saturday morning, he was immediately introduced to the possibility that the bubble might absorb enough oxygen to explode. The speculation was coming from Bethesda, and Denton was not prepared to say that it was wrong. However, he decided to check the accuracy of the story by calling in additional experts whom he trusted.

In the 8:00 to 9:00 A.M. range, Denton met with Met Ed President Walter Creitz or his representative on two occasions. During the first conversation, Denton asked Creitz to have a list of industry experts flown in to help analyze the bubble problem. There is no transcript of that conversation, and Creitz has never given any indication that he was in the least bit hesitant. However, when Denton talked to Hendrie shortly thereafter, he openly complained about Met Ed's lack of cooperation in bringing in the necessary experts.[29]

Before they finished, the White House operator broke in to say that the President wanted to speak to Harold Denton. Carter called because he was planning a trip later that day, and he wanted to be sure the reactor was in a state where he was not needed in Washington. Denton relayed the difficulties that he was having getting experts flown in. Carter promised to work on it. Carter had Jack Watson go over Creitz's head and call GPU President Herman Dieckamp. As Hendrie later understood from Dieckamp, Jack Watson really "read him the riot act, to get people down there."[30] Soon, Denton had all the help he wanted.[31]

The second call from Creitz related to the problems in the reactor. Until there was more than incomplete evidence that an explosion might eventually occur, Met Ed and Denton had to concentrate on the more immediate danger that the bubble might block the coolant flow. On that, there was finally some hint Saturday morning of good news. More precisely, Denton saw a hint, while John Herbein saw firm evidence.

Since the bubble had been unquestionably identified over twenty-four hours earlier, various attempts had been made to nibble away at it. As Mattson had been proclaiming to anyone who would listen, any attempt to play with the bubble might cause it to shift and block all coolant flow. Nevertheless, some tactics were available and had been tried.

For instance, with minor fluctuations in the primary loop pressures, it might be possible to expand the bubble low enough in the vessel that some would break off into the exit pipe to the hot leg. Small amounts could then be released through let-down valves or through the pressurizer to the containment building atmosphere. But too much of a pressure drop could break off too much of the bubble, causing a block in the hot leg. An even larger expansion could uncover part of the core.

With no level indicators in the core and with the "pneumatic volume" causing pressure indicators to stay at about 1000 psi however much water was drained out of the let-down system, there was no way to know how much fluctuation was too much or not enough. But noting that there were small releases of hydrogen escaping to the containment atmosphere, there appeared to be some venting of the bubble in progress.

Combined with this, the early NRC guess of 1000–1500 cubic feet of hydrogen at 1000 psi was refined on Saturday morning to 1000 cubic feet at 1000 psi. If one wanted to, one could call this a decrease in the bubble's volume by as much as a third. It was unusual arithmetic, but certainly not beyond the boundaries of some of the calculations in progress.

Creitz wanted to present this good news to the public, and he met with GPU's and Met Ed's press officers to plan a strategy. While they were talking, Joseph Hendrie called Creitz at Jody Powell's suggestion to recommend that GPU not release any more news stories that conflicted with Denton's information.[32] Because of this, Creitz sent GPU Vice-President William Murray to talk to Denton.

As he had made clear the previous day, Denton was firm that he would release no joint communiques with the company. Denton got Murray to agree that Denton should be giving the press conferences. However, this one had already been announced, and GPU decided to go ahead with it, even though they agreed that it would be their last.

At 11:00 A.M., on schedule, John Herbein addressed a press gathering at what Walter Creitz announced would be GPU's last press conference. Herbein did not acknowledge any possibility of an explosion in the core vessel, concentrating instead on how the bubble was shrinking in size and how there was no plausible danger of an explosion in the containment building. The question of an explosion in the bubble simply did not arise.[33]

Within a couple of hours of Herbein's "final" press conference, Hendrie held his press conference in Bethesda. The differences in content were dramatic, and Hendrie was repeatedly pushed to explain why Met Ed said the bubble was shrinking while he said it was growing and potentially explosive. He did not have a good answer and confused matters further by discussing the size of the bubble in the completely unrelated pressurizer.

Confusion and grumbling in the press were intense, and it was not long before the whole issue was brought to the attention of Jody Powell in the White House. The previous evening, Powell had instructed all interested parties that there were to be only three sources of official information on the crisis. Met Ed and Bethesda, neither of which was included on that list, had created an entirely new public crisis in confidence even though both had been specifically requested to avoid statements to the press. Jack Watson on Saturday afternoon called them both and made it as clear as possible that the White House expected them both to honor the President's request.[34]

The power of the White House to quiet Met Ed was questionable, but generally honored. Its power over the NRC was more direct. However, there were people in NRC who still did not get the message. About 9:00 P.M., Associated Press reporter Stan Benjamin called Edson Case and Frank Ingram at the Bethesda center to get an official reaction to a story he was releasing. He read them the exact text describing how tens of thousands of people might have to be evacuated before "risky operations" could be tried to vent the gas. If the operations failed, there could be a meltdown. If they were not tried, the bubble could explode within two days. Case and Ingram verified the story.[35]

Almost immediately, the story hit the wire. On hearing of it, Thornburgh in Harrisburg was enraged, because this type of story could cause panic and was not what he was hearing from the plant. He called the White House to get these releases stopped. A series of phone conversations began to the plant, where the White House found the NRC people to be both completely baffled and in disagreement with the report. When they finally tracked the story to Bethesda, Watson's assistant, Eugene Eidenberg, more or less ordered Case not to speak to the press for any reason.[36] Ingram received a separate call. By the time the calls had subsided, the White House had ordered the press center in Bethesda dismantled.

By requesting restraint, and then by issuing individual orders, and finally by dismantling the press center, the White House had corrected the almost continual flow of often inaccurate and inflammatory information from the well-intentioned engineers at Bethesda. It was now up to Harold Denton to correct the damage being done by the Associated Press story.

While he was doing that, evidence was mounting among the consultants that Bethesda might be right and Denton might be wrong. By Saturday evening, it was no longer clear that Denton had two days, or even two hours, before the reactor exploded.

Back at the reactor, Denton had spent Saturday fighting the bubble and informing the White House and the Governor of his progress. While no one else in the federal government seemed to accept the White House coordination plan as binding, Denton followed it rigorously. As a result, Denton's two contact agencies had enough information on the reactor to know when Denton felt that Bethesda's statements were unfounded.

Early in the day, Denton had requested a long list of industry representatives to be flown to the site to help fight the bubble. They were invaluable in two ways. First, those that Denton deemed qualified actually took over the operation of the controls. The legality of this is highly questionable, but the Met Ed employees were now exhausted and needed help.[37] Since several employees from other Babcock and Wilcox reactors had trained on the same simulator, Denton put them on the controls.

More importantly, Denton called in such experts as those who had done the hydrogen studies for the Rasmussen Report, which was the definitive NRC study on the consequences of accidents at reactors. He also called in any industry people who had experience with hydrogen overpressure.[38] By about 3:00 P.M. advisors whom Denton trusted told him that the entire scare was unfounded. While hydrogen could be ignited in a 5 percent oxygen mixture, it would not explode at these temperatures until the concentration was more than twice that figure. More important, at these pressures of hydrogen, oxygen would stay in solution in the water and not mix with the hydrogen. Rather, as happened during normal reactor operations, there was always free hydrogen suspended in the coolant that would absorb excess oxygen.[39]

In other words, a hydrogen explosion in the core was not possible. Denton and Vic Stello listened to the explanation and were convinced. Convincing Bethesda, where there was a different and actually more impressive set of experts, might prove to be a more difficult matter. In fact, it was getting more difficult all the time, as the evidence in favor of Bethesda's theory was growing.

Paul Critchlow in the Governor's office and Harold Denton spoke on the phone shortly after the Associated Press story was released at 9:00 P.M. They arrived at two specific plans to destroy the credibility of the explosion theory. First, Critchlow immediately issued a three paragraph press release. It quoted Denton and said "the report resulted erroneously from what he called a 'postulation' by engineers about the potential for the bubble. . . ." It said that Denton had learned by 3:00 P.M. that afternoon that the assumptions behind the postulation were incorrect and that there was no danger of an explosion.[40]

Second, they agreed that Denton should come to brief the Governor and that they should make a joint appearance before the press. Denton left for Harrisburg about 9:30 P.M. After about an hour, Thornburgh and Denton stepped before the cameras to make three major announcements. First, the reactor was safe from explosion. Earlier calculations had been based on erroneous assumptions.[41]

In fact, as a second announcement, Denton claimed that Bethesda agreed with him. The problem with the earlier Bethesda releases, Denton said, was that they were engineering scenarios that were presented as predictions. There was no disagreement between Denton and the Bethesda staff on what was actually happening. Instead, Denton said, "I guess it is the way things get presented."

Finally, to demonstrate just how safe things were, Thornburgh had just talked to President Carter. Thornburgh wanted to announce that the President and Rosalynn Carter were coming to Harrisburg tomorrow to tour the Three Mile Island plant. Obviously, if the President and First Lady were going to walk through the plant, the engineers must agree that the explosion theory was wrong.

But the engineers did not agree by any means. Back at Bethesda, Roger Mattson had gone home for the evening. But he left behind several experts who continued working on the problem. Into the early morning hours of Sunday, they could find no flaw in their calculations that an explosive mixture of hydrogen and oxygen was building.

Somewhere in the early morning hours, however, they began to agree that they had made one serious mistake in calculations. By their understanding, the explosive mixture that could rupture the core vessel was reached when the oxygen grew to 5 percent of the bubble. It would not self-detonate at that level, but any activity such as movement in the core would set it off.

By earlier calculations, the 5 percent mixture would not be reached for two or three days. Sometime after midnight, they realized that they had been too conservative in their figures. The oxygen level was already at 5 percent and climbing. It could explode and rip the entire nuclear system apart at any moment.

Later that morning, the President, the First Lady, and the Governor were all going to walk through a growing bomb that was destined to go off with the slightest disturbance. But their press center had just been rather brutally dismantled. If they let the slightest leak of this story get out, they would all certainly be fired. So they waited, and rechecked their figures, and checked them again. The expertise in that room was enormous, and there was no disagreement. Three Mile Island, with or without its special guests, was going to explode.

NOTES

1. Testimony of Harold Denton, President's Commission on the Accident at Three Mile Island, May 31, 1979, p. 321.

2. "Dewey Schneider's House," in "The TMI-2 Story," booklet prepared by General Public Utilities, May 25, 1979, pp. 12-13.

3. Testimony of Harold Denton, President's Commission, May 31, 1979, p. 321.

4. Met Ed news briefing, March 30, 1979, p. 11 (mimeographed).

5. "Report of the Public's Right to Information Task Force" to the President's Commission, October 1979, pp. 201-202.

6. Testimony of Harold Denton, President's Commission, May 31, 1979, p. 319.

7. Ibid., pp. 316-17.

8. Ibid., p. 313.

9. Testimony of Richard Thornburgh, Pennsylvania Select Committee on Three Mile Island, May 10, 1979, p. 15.

10. Stated during phone conversation in NRC meeting transcript, March 30, 1979, pp. 117-18.

11. NRC meeting transcript, March 30, 1979, p. 126.

12. Testimony of Harold Denton, President's Commission, May 31, 1979, p. 323.

13. Deposition of Harold Denton for President's Commission, Bethesda, Maryland, August 2, 1979, p. 120.

14. Testimony of Harold Denton, President's Commission, May 31, 1979, p. 324.

15. Deposition of Jessica Tuchman Mathews for President's Commission, Washington, D.C., August 23, 1979, pp. 61-62.

16. "Report of the Public's Right to Information Task Force," p. 201.

17. NRC meeting transcript, March 30, 1979, pp. 189–99.

18. NRC meeting transcript, March 30, 1979, p. 158.

19. NRC meeting transcript, March 31, 1979, section I, p. 40.

20. Testimony of Vic Stello, President's Commission, June 1, 1979, pp. 85–91.

21. NRC Preliminary Notification, March 31, 1979, 1:10 A.M.

22. Deposition of Roger Mattson for President's Commission, Bethesda, Maryland, August 6, 1979, pp. 181–83.

23. Joseph Hendrie speaking in NRC meeting transcript, March 31, 1979, section II, pp. 45–49.

24. Deposition of Roger Mattson, pp. 185–86.

25. Ibid., pp. 190–96.

26. Deposition of Joseph Hendrie for President's Commission, Washington, D.C., September 7, 1979, p. 218. Testimony of Joseph Hendrie, President's Commission, April 26, 1979, p. 131.

27. "Report of the Public's Right to Information Task Force," p. 210.

28. Testimony of Harold Denton, President's Commission, May 31, 1979, pp. 344–45.

29. NRC meeting transcript, March 31, 1979, section II, pp. 6–9.

30. Ibid., p. 26.

31. Deposition of Jack Watson for President's Commission, Washington, D.C., September 6, 1979, pp. 104–105.

32. Deposition of Joseph Hendrie, p. 217.

33. Met Ed news briefing, March 31, 1979, p. 4 (mimeographed).

34. Deposition of Jack Watson, pp. 90–91.

35. NRC meeting transcript, March 31, 1979, section II, p. 40.

36. Deposition of Jack Watson, p. 93.

37. Testimony of Harold Denton, President's Commission, May 31, 1979, p. 311.

38. Ibid., pp. 361–62.

39. *Report of the President's Commission on the Accident at Three Mile Island* (Washington, D.C.: U.S. Government Printing Office, October 1979), p. 133.

40. Press Release 323-D79, Governor Thornburgh's Press Office, March 31, 1979.

41. Press Conference of Richard Thornburgh and Harold Denton, Harrisburg, Pennsylvania, March 31, 1979, p. 11.

Chapter 12

Upstairs, Downstairs

Had all gone according to plan, information about the reactor would have been transferred from Met Ed to the Bureau of Radiation Protection to PEMA to the Lieutenant Governor. That plan worked until Wednesday afternoon, when the event became too exciting and too important to leave to administrators.

Had all gone according to plan, information about evacuation preparedness and response would have flowed up and down the levels of command in PEMA. Until Friday, except for the unwelcome company of a couple of intruders, it did. But by Friday the possibility of evacuation became too real, and the planning became too exciting. On Thursday Colonel Henderson heard the first reference to a ten mile evacuation. By Friday, Mattson in NRC was discussing at least twenty miles. By Saturday, PEMA was receiving orders, but had been excluded from gubernatorial meetings.[1] The political heavyweights had taken over. As Margaret Reilly stated the problem: "For years they [NRC] had been preaching to the states that offsite consequence management is yours, that is not NRC, it is yours. Then it came down to being ours and they just couldn't let it alone (*sic*)."[2]

One of those least inclined to leave it alone was Joseph Califano at HEW. As already stated, Califano was genuinely concerned that evacuation decisions were left to such non-health-related agencies as the NRC and DOE. It apparently did not occur to him that evacuation decisions could be made only by the Governor and that the Governor might have his own advice machinery. But then, Califano and almost all other active federal officials were also unaware that they were signatories to an agreement to supplement state needs.

Instead, at the Wednesday evening meeting of health-related agencies at HEW headquarters, Califano proceeded with his own health-related planning and incorporated it into a memo dated the following day.[3] Under this plan, the Food and Drug Administration would intensify its food and water sampling in the Harrisburg area. They also gave the state 250 dosimeters to aid the Bureau of Radiation Protection in off-site monitoring.[4] HEW personnel would be placed in NRC's incident response action center and at DOE's center at the Capitol City Airport in Harrisburg. The Public Health Service would immediately concentrate on training about thirty of its physicians to treat radiological injuries. This was done without checking to see if Pennsylvania needed them, which they did not.[5] HEW was also unaware that DOE was doing the same thing.

On the issue of advising an evacuation to the President, Califano held off for one more day. Of course, the President had no powers in this area. However, since the White House had insisted Friday afternoon that all communications were to go through them, they were the only group that Califano could advise.

However, the most time-consuming activity that Califano began on Friday was a search for potassium iodide. During a radiological accident, the gasses floating through the air in the highest concentrations are isotopes of xenon and iodine. Compared to most radioactive elements, xenon is not a severe threat to the affected population. Its decay products are relatively mild. Chemically, it reacts with nothing. Therefore, it dissipates quickly and can be washed off easily.

Radioactive iodine is a different matter. Exposed to the skin, it is also relatively harmless. But if it is inhaled, it sticks in the thyroid gland, where it continuously bombards some of the body's most vulnerable areas with radioactivity. Radioactive iodine is such a potential health hazard that special iodine filters are installed in nuclear plants to try to restrict the uncontrolled accidental releases to xenon.

There is a medical defense against radioactive iodine poisoning that has been known to be effective for perhaps twenty years. If a person takes a dose of supersaturated potassium iodide before exposure, it literally clogs the thyroid gland. Radioactive iodine can then be inhaled and will not be absorbed into the thyroid. It is then normally exhaled before it can cause damage.

Given that the best defense is not to inhale iodine, potassium iodide is an extremely effective alternative that seldom has serious side effects. As a result, the Food and Drug Administration approved it for emergency use by the general public in December 1978.[6]

FDA Director John Villforth at first thought that the drug could be purchased from Harrisburg pharmacies.[7] When FDA began checking inventories of the drug in the Harrisburg area on Friday evening, however, they found that the dosages on hand were insignificant. The situation was not expected to be better in Baltimore, Philadelphia, or other nearby cities, and the FDA was concerned that a middle of the night inventory of drug stores would cause a panic. While some of the drug existed to cover the scattering of medical conditions for which it is appropriate, quantities sufficient to fight a general radiological release were not stockpiled anywhere since pharmaceutical firms had never produced them.[8]

FDA had a problem. The drug could not be manufactured under normal conditions until long after it might be needed. There were pharmaceutical firms that could go into emergency production. However, a purchase order and authorization through HEW could take weeks. John Villforth at FDA was on the spot. He understood that traces of iodine were already being recorded off site, but bureaucratic procedures prohibited him from buying the needed drug on a timely basis.[9]

Finally, about 3:00 A.M. Saturday morning, Jerome Halperin within FDA's Bureau of Drugs reached a solution. Mallinckrodt Chemical Company in St. Louis would go into around the clock emergency production of potassium iodide without a work order. Parke-Davis in Detroit joined in the effort, as did a medicinal dropper company in New Jersey. FDA's offices in Rockville, Maryland, began printing instructions and bottle labels.

By 1:30 A.M. Sunday, less than one day after the agreement was reached, the first bottles reached Harrisburg. Within four days, the full order of 237,013 bottles was on site. But it still was not clear who would pay for it or who would decide how and when to distribute it.

Another federal agency was having far more difficulty responding to the crisis. Douglas Costle at EPA had ordered his monitoring equipment shipped from military duty near Las Vegas to Harrisburg after Friday's scare. They had seventeen people and a considerable amount of delicate equipment. Late Friday evening, they began making arrangements for the transfer.

However, United Airlines had a strike scheduled for midnight, and commercial air travel was hopelessly overbooked. They tried to charter an aircraft, but none could get to Las Vegas for at least twenty-four hours. Finally, they explained the crisis to TWA, who bumped several passengers to get the equipment from Las Vegas to Philadelphia. They could get it no closer than that by air, and EPA people in Philadelphia had to drive it the rest of the way.

The personnel and the rest of the equipment were put on a commercial flight to Chicago. From there, the only way to get to Harrisburg was by making a connection all the way back in Los Angeles. They flew to Los Angeles. However, one person was then bumped in Los Angeles and did not get to Harrisburg until much later.[10] The rest arrived about noon Saturday.

By noon Saturday, events in Washington, D.C., were escalating. Califano called a meeting of his "health cabinet" to check on their progress and to get back to the evacuation question. With all the conflicting information floating around, they were still unsure what to recommend. However, it bothered them deeply that the NRC was becoming the only permitted channel of information. They had the distinct impression from NRC's answers at the Friday meeting that NRC had little concern with health physics problems.[11]

They took three actions. First, Califano called Carter to ask that a limited cabinet meeting be called. Carter, who was preparing to leave town for the evening, passed the call to Jack Watson. Watson vetoed the idea of a cabinet meeting because the press would blow the significance of it out of proportion.[12] However, Watson was willing to meet with second-ranking officials, which he did about 5:30 P.M. that afternoon.

Second, the health officials in Califano's office discussed the possibility of an evacuation. Since the NRC had left the impression on Friday evening that a cold shutdown of the reactor was not close, everyone in the room began advocating at least a partial evacuation. The five mile evacuation idea was quickly dismissed. The debate was soon between FDA's John Villforth, who wanted a ten mile evacuation, and the National Cancer Institute's Arthur Upton, who wanted twenty.

Finally, Califano drafted a memo that was delivered to Jack Watson about noon. It argued that the President should recommend to Thornburgh that an evacuation begin immediately unless the NRC could assure a cold shutdown. The exact distance was not stated, although Califano felt that the population up to twenty miles away should be notified "publicly and officially" that they might have to evacuate on six hours notice.

Second, Califano argued that the NRC should not be permitted to intervene into the reactor's operations without checking with

HEW and EPA on health consequences. Finally, Califano outlined what HEW was already doing and asked what else they might do.[13]

When Jack Watson received the memo Saturday afternoon, he called Harold Denton to get his impressions. Denton dismissed it as unnecessary. So did Jessica Mathews, Science Advisor Frank Press, and Gene Eidenburg.[14] The message was not passed to President Carter or Governor Thornburgh since it did not seem to deserve that attention. Instead, Watson's updating memo to Carter mentioned only that Califano's request for a cabinet meeting had been rejected in favor of a staff meeting.

HEW had one more chance to impress those at the top with what they saw as a dangerous policy of ignoring the health experts. During the afternoon, Califano drilled Richard Cotton, Arthur Upton, Anthony Robbins, and John Villforth on the presentation that they would make to the White House meeting that evening.[15] As they left for the 5:00 P.M. meeting, they were prepared and duly psyched to take the meeting by storm.

But one does not easily take Jack Watson by storm. To begin, the meeting started almost thirty minutes late, which should have surprised no one in a government meeting, but which did not help the sense of urgency or momentum in the meeting. Then Watson began the meeting with the sobering announcement that the meeting was intended only to exchange information and not to make decisions. The contents should be considered confidential.[16]

Watson then reemphasized that the federal government *must* maintain a low profile because press statements from a variety of sources were alarming the public and violating agreements with the state. Instead, all information needed to be passed through the White House to the NRC.

This was not what HEW needed to hear, because Richard Cotton now needed to argue that health-related information was being suppressed from the public by the NRC. As he started, Watson quickly cut him off by turning to NRC Commissioner Victor Gilinsky and saying "please take care of that."[17] The NRC, which HEW saw as being the problem, was being asked to fix it.

The meeting progressed to evacuation plans, and FDAA reported that a ten mile evacuation could be completed in four hours while a twenty mile evacuation could be completed in five hours. By state calculations, where the realities of evacuation were much closer to home, these estimates would have been considered ludicrous. However, the state was not at the meeting.

The topic gave Richard Cotton one more chance to say that the people in that zone should be alerted. Watson cut him off again, saying that he had read Califano's memo and that HEW should send people to NRC's incident response action center. Of course, they were already there.

Then Watson introduced another matter that helped to fan an existing confusion. The Department of Energy had been at Harrisburg monitoring off-site releases under Joe Deal since the morning of the accident. They claimed the auspices of the IRAP in their press releases. However, they were left out of Watson's Saturday meeting simply because Watson's staff forgot to invite them.[18]

On Friday, DOE had agreed with Pennsylvania's Bureau of Radiation Protection to coordinate all off-site monitoring. At Watson's meeting on Saturday, only Villforth from FDA was apparently aware of the DOE's role or even of the IRAP. Villforth had long since decided that the provisions of the IRAP were being completely ignored.[19] Therefore, Watson assigned the NRC to coordinate off-site monitoring.[20] There was no objection in the room.

The strange saga of NRC and DOE continued from Wednesday through Sunday. Both had been monitoring off-site radiation extensively since Wednesday morning. Both had been sharing their information on a regular basis. Yet neither was aware that the other's efforts in this activity were sizable. In addition, both agencies thought that the other realized that their own agency was in charge. Just to complicate things, after the Saturday meeting, both apparently were in charge.

On Sunday, that situation finally broke open. NRC discovered that DOE had been collecting radiological information that NRC needed. When NRC made this alleged intrusion known at the White House, HEW and EPA found out. Now HEW and EPA went back to Eugene Eidenberg at the White House to complain that the government's official proponent of nuclear engineering should not be doing the monitoring. Eidenberg proposed another memo to Watson.

HEW and EPA's argument was convincing. However, Watson was becoming tired of their attempts to continually increase their role in the recovery effort. On April 13, long after the crisis of Three Mile Island was over, Watson finally switched the responsibility for off-site monitoring to EPA.

Early Thursday evening, March 29, things were relatively quiet in the emergency management machinery in Pennsylvania. The reactor seemed stable, and the periodic update from PEMA to the local emergency management directors was a rather uninformative "no change." Kevin Molloy in Dauphin County was pushing his volunteer directors beyond the bounds of friendly persuasion. He was convinced that their plans were virtually useless, and he wanted them improved.[21] He was also in periodic contact with the Red Cross and the local ham radio operators, although there was little to tell them.

It was difficult to generate any excitement in the evacuation machinery. Many people had left the area on their own, although the lack of anxiety was demonstrated by the unusual statistic that automobile accidents were lower than would have been expected with no crisis.[22] The Pennsylvania State University campus at Middletown was still open, as were all public schools.

Ironically, while the highest off-site releases of the accident were on Thursday evening, they were peaking at a merely bothersome 10 millirems per hour. Other sampling further off site showed nothing. As part of their normal sampling schedule, the state of Maryland just across the border happened to draw air and water samples on Wednesday, the first day of the accident. They found background levels of radiation.[23]

The Pennsylvania Department of Agriculture began sampling milk from local farmers on Thursday, and they found almost nothing.[24] They did, however, provide a lucrative temporary scheme for con artists. Some people, posing as agricultural inspectors, showed up at local farms asking for gallons of milk to take to the lab for testing. Once discovered, a radio announcement was sufficient to put an end to the free milk.

But the difficulties began on Thursday night, when Colonel Henderson heard the first NRC estimate that the reactor was in trouble and might need a ten mile evacuation. The state was ready for five miles and could have evacuated that radius within about six hours.[25] But a ten mile plan eliminated most of the relocation centers for people in the five mile zone.

Thursday night, Henderson called the county directors for Dauphin, York, and Lancaster counties and ordered them to begin working on ten mile plans. This presented no particular problem to York County since almost everyone within ten miles was in Goldsboro or Newberry townships and plans for those already existed.[26] In fact, about half of the population in those townships had already left.[27] Similarly, the new radius caused little difficulty for Lancaster County.

In Dauphin County, Molloy was more concerned. He spent the evening and much of the night trying on his own to plan for a ten mile evacuation. Progress was slow since there were still no visible signs to stir the evacuation machinery into action.

The signs came Friday morning. First, there was Molloy's call from what seemed to be a panicky Jim Floyd at the plant. Then Colonel Henderson called the three affected counties to say that an evacuation order was imminent. That order did not come, but the emergency management directors next heard the Governor's office issue an advisory to stay indoors within ten miles. Within three hours, the schools in a five mile radius were closed, and children and pregnant women were asked to leave. The Red Cross and PEMA half-borrowed and half-seized the Hershey Sports Arena as a shelter.

Suddenly, there were no more skeptics in the state evacuation machinery. But now that they took their jobs seriously, their plans were useless. Colonel Henderson passed along the message that he had picked up from Harold Collins that the pending evacuation would be for ten miles. There were no such plans on the books.

Within five miles of the plant, the population is sparse and easy to move. In fact, on Friday most of the local population moved on its own. But the next concentric circle adds over 100,000 people, three major hospitals, and twenty nursing homes.[28] Pennsylvania was not prepared for this, and all levels of the emergency management structure needed help.

Some help came without asking. Pennsylvania State University in Middletown closed at 2:00 P.M. and sent its students home, even though it was outside the five mile limit.[29] Hershey hospital, eight and one-half miles away, began discharging all the patients that it safely could.[30] Several other hospitals followed the lead. By Saturday, there were only about sixty people left in Goldsboro.[31]

Some help volunteered itself. For instance, when the Defense Civil Preparedness Agency called Colonel Henderson again after being rebuked on Wednesday, Henderson accepted the offer of evacuation planners to be sent to each of the affected counties. This was particularly useful in places like Franklin County, that were removed from the reactor, but that suddenly found themselves to be potential hosts to evacuees.[32] Help also came from places such as the Food and Drug Administration, which began work on the vials of potassium iodide. While these vials were first destined for storage by the Bureau of Radiation Protection, they ended in the possession of PEMA.

But much of the help was specifically requested from other parts of the state. For instance, the Pennsylvania Department of Transportation was able to find 300 vehicles, fully fueled.[33] Most were personnel-carrying trucks. Kevin Molloy was assigned 122 state police and 800 National Guard troops to help keep order and staff the potential evacuation.[34] With the swamp of calls, all county offices requested and received extra phone lines.[35]

But some of the help requested or expected from the state did not come. Plans for evacuation beyond the five mile radius absolutely required that all available school busses be pressed into service. Escape plans were designed using that assumption. But then the State Department of Education refused to close schools on Monday further than five miles from the reactor, and the principal transportation assumptions behind the evacuation plans collapsed.[36]

Before the ten mile plans were even partially devised, Harold Denton told PEMA Saturday night that the evacuation radius would have to be twenty miles. By this time, it was becoming clear that Colonel Henderson was merely receiving such decisions rather than sharing in them. Also, federal offers and intrusions of assistance were not systematically passing through some office where they could be easily catalogued to help with the planning.

Jack Watson in the White House assigned FDAA's Robert Adamcik to be the federal coordinator on site. However, Adamcik was specifically told to talk to the Governor and not to bypass the Governor by reaching into the state bureaucracy or the press.[37] Also, the President and the Governor agreed that the public relations would be better if Thornburgh did not request a state of emergency.[38] The net effect of this was to reduce the emergency powers available to tie the local efforts to Adamcik's office.

However, the emergency planning obviously continued after 8:30 P.M. Saturday when the radius was expanded to twenty miles. Now there were 700,000 people to evacuate, including the state capitol and the records in most state offices. There were also thirteen major hospitals and a large prison.[39] There were too many nursing homes to count. Nowhere nearby were there any large cities that could absorb this many people. Just to complicate matters, the plans would have to be hammered out on Saturday night and Sunday, when many facilities were not staffed.

Then there was the inevitable question of what to do beyond twenty miles. The city of Lancaster is twenty-three miles away. As a result, they did not receive notification to evacuate.[40] Instead, the lodging facilities that they had available as an Amish tourist area were a natural resource for storing evacuees.

But Lancaster is directly down river from Three Mile Island, and the city leadership was not going to sit back simply because they were three miles beyond the radius. Therefore, Lancaster quickly began its own independent evacuation planning, and they notified Lancaster County of what they were doing.

Lancaster County had considerable numbers of people within a twenty mile radius, and they had the help of PEMA and the federal officials from DCPA. Together, these groups put together the most sophisticated local coordination efforts in the entire crisis.

There simply are not many ways to leave the Three Mile Island area by car. From the Harrisburg area, travelers can leave on the Pennsylvania Turnpike, although they would have to approach to within almost five miles of the plant to get on the turnpike. They could travel north on Interstate 83, but traveling south would take them within two miles of the plant.

On the southern side in York and Lancaster counties, the options were more limited. Interstate 83 had two lanes going south to Baltimore. There was also a state four lane highway intersecting Interstate 83 across the river from Lancaster City.

If the southern tier had to be evacuated, some very rural areas were going to find themselves suddenly being host to a massive number of people. Therefore, Lancaster County called a meeting for 2:00 P.M. on Sunday to work out the details.[41] The meeting was attended by the state police (who would have to block roads to be one way), the Red Cross, and representatives from Chester, Bucks, Montgomery, and Philadelphia counties in Pennsylvania and Cecil County in Maryland.

At that meeting needs were assessed, and Paul Leese from Lancaster County agreed to go to PEMA headquarters to get more portable radios. Distant counties were asked to find shelters, and affected counties were asked to concentrate on locating invalids.

Some needs, particularly for relocation space, simply could not be met. Therefore, the next day, a representative of the Maryland Civil Defense was brought in for a briefing and to integrate his resources into the planning effort. The same thing was done with Franklin County, Pennsylvania. Franklin County already had extraordinary evacuation plans and relocation sites in place since it is the home of "Site R," the underground Pentagon to be used in case of nuclear attack.[42]

Back in Dauphin County, Molloy helped two nursing homes evacuate on Saturday as a precautionary measure.[43] The difficulties of that experience taught him what Lancaster was learning at a more fundamental level—wide-scale rapid evacuation simply was not going to work.

Nevertheless, plans continued. By Sunday, hospital patients in Dauphin County were reduced from a high of 3000 down to 1300.[44] The Department of Health was beginning to stockpile supplies of the arriving potassium iodide. By a technique of moral persuasion reserved for extraordinary times, Molloy got Amtrak to park a northbound train in Harrisburg in case it would be needed. Still, over the weekend, an evacuation beyond five miles would have required days instead of hours.[45]

It did not help that the emergency management telephone numbers, which even the mayors forgot on Wednesday, seemed to be

public knowledge by the weekend. Placing calls was an achievement
with all the inquiries coming in. But the emergency management
people got their information from PEMA, and PEMA knew next to
nothing. What is now accepted as the classic PEMA communique
reached the counties at 10:45 P.M. on Saturday. It read, "NBC
broadcast that the bubble burst or bubble growing and that mass
evacuation occurring was spurious."[46]

That message created the storm that finally caused public communications to be moved elsewhere. State Senator Gekas happened
to be in the Dauphin County office when that teletype message
arrived. He called the Governor and Lieutenant Governor to get a
clarification. Both were too busy to speak to him.

So Gekas dropped a bombshell. He told the Lieutenant Governor's
office that if Dauphin County did not get better information soon, it
would begin evacuating the entire county at 9:00 A.M. the next
morning.[47] At the point, the people in Dauphin County were not
sure whether they intended to carry out the threat. But it apparently
had its impact with Scranton.

At 2:00 A.M. Sunday, Scranton called the Dauphin County center
to calm them down. They were adamant that evacuation would start
in seven hours. At 8:00 A.M. Scranton sent Henderson to talk to
them, but they soon had Henderson convinced that he could not
help them either. As they talked, the 9:00 A.M. deadline passed. But
Henderson went back to Scranton to say the problem had not been
resolved.

Finally, at 10:00 A.M., Scranton himself went to the center. When
he got there, according to Molloy, "I think he was just totally
shocked by what was transpiring at our level; how busy we were;
how much work we were doing; how complicated it was."[48] Scranton promised to find some way to distribute the work load, and the
immediate crisis ended. The method they found, however, had its
own limitations.

Within Pennsylvania, there is a Governor's Action Center, which is
a complaint and research bureau for the citizens of the state. If a
citizen feels that his or her water has been cut off unjustly, or a
welfare check is missing, or a local jurisdiction will not clean a storm
drain, he or she can call the number. Researchers, or "advocates,"
check into the story and get back to the complainant with an answer.
The staff has nothing to do with nuclear engineering, but it is professional in terms of restraint, research, and rumor control.

The emergency management machinery was swamped with calls
that were trivial to their purposes but vital to the citizens involved.
Was it safe to wear clothes dried on a line? Should seven year olds be
evacuated? Was it safe to use cattle feed stored outdoors? Was it

safe to leave pet dogs in the yard? These were questions ideally suited for the Governor's Action Center, once they were given a list of sources to find answers.

The center was closed for the weekend. Times were hard in Harrisburg, and when the center opened Monday morning only eight employees arrived.[49] Because of the limited staff and the obvious crisis brewing outside, Director Charles Kennedy and his staff agreed that only TMI-related calls would be taken. Through the morning, they mainly collected questions that they would have to research and call back to answer.

Just after noon, Kennedy was called to the Governor's office to talk to PEMA, FDAA, and telecommunications people to see if his office could take some of the load off emergency management phone lines. Kennedy convinced them that he did not have the staff to accept more phones, but that his office was already doing rumor control. The PEMA people gave Kennedy phone numbers of such offices as the NRC to research questions and he agreed that he would take on this role for the state.

It took the Governor's Action Center almost a full day to get enough people assembled to go on a round the clock operating schedule. The biggest problem was not answering the phones, since supplemental people were quickly found. The center was also able to get information from the NRC that was reasonably accurate and timely.

The problem was that no one at the center could understand the information. When the NRC gave the center its first off-site reading in picocuries per cubic meter, no one at the center knew what they were talking about. It was hours before someone at the center remembered that he had a distant relative who taught nuclear physics at the University of Pittsburgh. That poor soul became the official interpreter over the next few days of any strange term that wandered through the center during the day or night.

At 9:40 P.M. Tuesday, Governor Thornburgh announced on radio that rumors and questions could be checked through the center, for which he gave the number. Until well into the morning, no line stayed open at the center long enough to make a call out.

The diversion of phone calls helped the emergency management machinery some. But what helped them most was time. By Wednesday of the second week, Paul Leese in Lancaster was able to call a meeting of all the affected counties in Pennsylvania and Maryland to compare notes.[50] The days that they would have needed over the weekend had transpired.

As a bottom line, given the six to eight hours notice that the evacuation machinery would have had, most of the twenty mile radius could have been cleared on Wednesday, seven days after the accident. There would have been some problems with the Amish, who accept these things as fate. But the real crisis would have been in Harrisburg. Had a complete meltdown been so cooperative as to have waited until Wednesday, much of Harrisburg would simply have been left behind.

NOTES

1. *Report of the President's Commission on the Accident at Three Mile Island* (Washington, D.C.: USGPO, October 1979), p. 41.

2. Quoted in "Report of the Office of Chief Counsel on Emergency Response" to President's Commission, October 1979, p. 63.

3. Deposition of Richard Cotton for President's Commission, Washington, D.C., August 16, 1979, pp. 28–29.

4. "Technical Staff Analysis Report on Public Health and Epidemiology" for President's Commission, October 1979, p. 73.

5. Deposition of Anthony Robbins for President's Commission, Washington, D.C., July 27, 1979, p. 47.

6. *Report of the President's Commission*, p. 41.

7. "Report of the Office of Chief Counsel on Emergency Response," p. 91.

8. "Technical Staff Analysis Report on Public Health and Epidemiology," p. 73.

9. Testimony of John Villforth, President's Commission, August 2, 1979, p. 274.

10. "Report of the Office of Chief Counsel on Emergency Response," p. 90.

11. Deposition of Richard Cotton, pp. 62–64.

12. Deposition of Jack Watson for President's Commission, Washington, D.C., September 6, 1979, pp. 66–67.

13. Content described in "Report of the Office of Chief Counsel on Emergency Response," pp. 92–97.

14. Deposition of Jack Watson, pp. 62–63.

15. Deposition of Richard Cotton, p. 49.

16. Ibid., p. 55.

17. Ibid., p. 57.

18. Deposition of Stephen Gage for President's Commission, Washington, D.C., August 13, 1979, p. 63.

19. "Report of the Office of Chief Counsel on Emergency Response," p. 108.

20. Deposition of Stephen Gage, p. 74.

21. Testimony of Kevin Molloy, President's Commission, August 2, 1979, pp. 7–8.

22. Testimony of Kevin Molloy, Pennsylvania Select Committee on Three Mile Island, Pennsylvania Legislature, July 25, 1979, p. 4.

23. Personal interview with Robert Corcoran, Baltimore, Maryland, June 14, 1979.

24. Testimony of Secretary Penrose Hallowell, Pennsylvania Select Committee on Three Mile Island, August 21, 1979, pp. 71-72.

25. Testimony of Kevin Molloy, President's Commission, August 2, 1979, p. 30.

26. Testimony of Leslie Jackson, Pennsylvania Select Committee on Three Mile Island, August 7, 1979, p. 80.

27. Testimony of Bruce Smith, President's Commission, May 19, 1979, p. 268.

28. Deposition of Oran Henderson for President's Commission, Harrisburg, Pennsylvania, July 30, 1979, p. 65.

29. Testimony of Theodore Gross, President's Commission, May 19, 1979, p. 21.

30. Testimony of Kenneth Miller, President's Commission, May 19, 1979, p. 114.

31. Testimony of Ken Myers, President's Commission, May 19, 1979, p. 37.

32. Testimony of Jere Gonder, Pennsylvania Select Committee on Three Mile Island, August 8, 1979, p. 5.

33. Testimony of Thomas Larson, Pennsylvania Select Committee on Three Mile Island, August 22, 1979, pp. 99-101.

34. Testimony of Kevin Molloy, Pennsylvania Select Committee on Three Mile Island, July 25, 1979, pp. 35-37.

35. Testimony of Paul Leese, Pennsylvania Select Committee on Three Mile Island, July 24, 1979, p. 9.

36. Testimony of Kevin Molloy, Pennsylvania Select Committee on Three Mile Island, July 25, 1979, pp. 20-21.

37. Testimony of Robert Adamcik, President's Commission, April 26, 1979, p. 39.

38. Testimony of William Wilcox, President's Commission, April 26, 1979, p. 13.

39. Testimony of Oran Henderson, President's Commission, August 2, 1979, p. 46.

40. Testimony of Albert Wohlsen, President's Commission, May 19, 1979, p. 52.

41. Details from testimony of Paul Leese, Pennsylvania Select Committee on Three Mile Island, July 24, 1979, pp. 10-15.

42. Testimony of Jere Gonder, Pennsylvania Select Committee on Three Mile Island, August 8, 1979, p. 35.

43. Deposition of Kevin Molloy for President's Commission, Harrisburg, Pennsylvania, July 26, 1979, pp. 65-67.

44. Ibid., pp. 65-69.

45. "Report of the Office of Chief Counsel on Emergency Response," p. 101.

46. Deposition of Kevin Molloy, p. 85.

47. Ibid., pp. 88-89.

48. Ibid., p. 100.

49. Details taken from "Establishing a Viable Public Information Center During Crisis Situations," Governor's Action Center, Harrisburg, Pennsylvania, May 7, 1979 (mimeographed); and personal interview with Director Charles Kennedy, Harrisburg, Pennsylvania, June 18, 1979.

50. Testimony of Paul Leese, Pennsylvania Select Committee on Three Mile Island, July 24, 1979, p. 43.

※ *Chapter 13*

Not With A Bang

When Andrew Jackson was president, people who wanted to see him just walked up to the front door of the White House and knocked. If he was at home, he often answered personally. People simply did not shoot presidents in the 1820s, and the sense of security around the White House was much less obvious than it is in private residences in Washington today.

But by 1979 times had obviously changed. When the president planned a trip, days and often weeks of preparations went into the scheduling and security arrangements. Presidents were still threatened and occasionally even killed. But with the heightened awareness in the Secret Service that any public exposure of the president was potentially dangerous, the chances taken by presidents were not as great as they were through most of the nation's history.

Despite all this, on March 31, 1979, President Carter agreed with Governor Thornburgh that the best interest of the citizens would be served if the two were seen within the next couple of days touring the Three Mile Island nuclear plant. While the plant was far from the end of its crisis, the two decision-makers wanted to demonstrate by their visit that the situation at the plant was not as dangerous as some of the less restrained press reports had made it sound. Accordingly, arrangements began for a presidential tour, complete with Governor Thornburgh and Rosalynn Carter, to be conducted during the afternoon of Sunday, April 1.

In some circles, such as the staff of the National Security Council, sudden preparations and odd assignments in the middle of the night were almost taken for granted. Zbigniew Brzezinski called Jessica

Mathews at home at 2:00 A.M. Sunday to ask her to go in to work early and to prepare a memorandum on the state of the reactor.[1] On completion the memo was to be passed to Brzezinski, who would pass it to the President.

Mathews went to her office to begin compiling her memorandum. To gather the information she needed, she called Harold Denton at the plant.[2] He was, after all, the designated official source of government information on the state of the reactor. Given the past performance of the team at Bethesda, that designation had been a wise choice. However, on the morning of April 1, the information that Mathews collected from Denton was more optimistic than the majority view among the experts on the explosive potential of the reactor.

During the early morning hours of Sunday, calculations and speculations related to oxygen generation and potential explosions in the reactor had been the center of attention of two separate groups of experts. The most active group was in Bethesda. Roger Mattson went home to get some sleep at about the same time that Mathews was being called to write a summary for the President. As he left, the best guess at Bethesda was that the hydrogen bubble would reach flammable limits within two to three days.[3]

But shortly after Mattson left, the staff members still working on the problem came to the sobering realization that they had been underestimating the rate of oxygen generation in the core vessel. Instead of the core being within two to three days of a flammable mixture of 5 percent oxygen, it had already reached 5 percent and was climbing.[4] At any moment, the bubble in the core vessel could explode.

It was a tense situation, made much worse by the realization that the President planned to tour the plant in a matter of several hours. Something clearly needed to be done quickly, and the team at Bethesda began working in several directions.

First, they notified Harold Denton. He had his own team of experts—the industry group in "Trailer City" behind the observation center now numbered in the hundreds.[5] He assigned Vic Stello to gather a team and to concentrate on the new calculations received from Bethesda.

Stello's team, which was heavily weighted toward industry representatives from such places as Westinghouse and Babcock and Wilcox, worked throughout the night. By early morning, they reached the conclusion that there were two flaws in Bethesda's calculations.

First, the industry representatives who had experience with hydrogen overpressure felt that the oxygen being formed could not force itself into the hydrogen bubble the way that Bethesda assumed. The formulas and the thrust of theoretical nuclear physics were on the side of the Bethesda group.[6] However, the industry representatives working with Stello felt that experience had not shown those formulas to be correct.

Second, the group under Stello felt that the Bethesda group had made a fundamental and rather elementary calculating error. Oxygen generation occurs continuously in reactors even during normal operation. To counteract this, hydrogen gas trapped in the top of the makeup tank slowly flows through the core vessel and recombines on its own with the oxygen. While the exact figures on this process were not known, Stello's group suspected that the oxygen level was not building at all.[7] At the very least, Stello's group was baffled by the assumption at Bethesda that no natural recombination was occurring.

By the time that Mathews called Denton to get information to pass to the President, Denton was convinced from Stello's figures that the reactor was not dangerous. Denton told Mathews that he felt better about the state of the reactor than he had the day before.[8] He warned her that he still knew no completely safe way to vent the bubble out of the core vessel, and that any pressure or coolant flow malfunctions while the bubble was still there could be potentially catastrophic. However, there was no evidence of deterioration. In fact, it even appeared from preliminary calculations that the bubble had been reduced from about 1000 cubic feet to about 800 cubic feet.

When Mathews put the information into a memo and passed it to Brzezinski, the two discussed the implications of the President's visit. They did not consider the reactor or the President to be in any immediate danger. Rather, they understood that the risks would begin when the operators tried to vent the hydrogen bubble. That eventual venting introduced two concerns for the presidential advisors.

First, they were worried that the President might get asked by the press about the risks involved in trying to reach a cold shutdown. The risks were enormous, and an honest answer could destroy any reassuring effects that the presidential visit might have. Therefore, Brzezinski and Mathews agreed to recommend to the President that he avoid commenting at all on the physical condition of the reactor.[9]

The second concern was raised when Mathews called Vice-President Walter Mondale to brief him on the reactor and on the President's visit. While there was disagreement among the experts on the current stability of the reactor, everyone agreed that the process of venting the bubble was loaded with potential trouble. Many people, particularly those at the NRC in Bethesda, were in favor of a precautionary evacuation before such a process was started. The best guess from Pennsylvania was that even a five mile evacuation would take about four hours.[10] Should an explosion rupture the primary coolant loop, astronomical off-site releases could begin in as little as thirty minutes. Mondale's major concern that he expressed to Mathews was that the government should err in favor of such an evacuation rather than gamble that everything at the plant would go as planned.[11]

The memo was then passed to Jody Powell and Jack Watson, and both discussed the contents with Jessica Mathews. After rewriting, it was passed to the President.

In the memo, Carter was told of the dangers of venting the bubble and the strong possibility that a precautionary evacuation might be needed.[12] He was specifically requested not to comment to the press on the condition of the reactor. He was also advised that he should make it clear to the population that the crisis was not over and that the people might still be asked to evacuate as a safety precaution. The memo had been approved by all the relevant advisors in the White House machinery. Based on his later actions, Carter seems to have accepted the technical summary at face value.

But as the morning progressed, other members of the federal establishment were frantically trying to convince those making the decisions that there was imminent danger of an explosion. Roger Mattson and a few senior staff members met with three of the NRC commissioners at 9:00 A.M. Sunday. They restated and agreed to their findings that the hydrogen bubble contained 5 percent oxygen and was increasing at a rate of about 1 percent a day. The bubble was already potentially explosive, although it was still a few days from spontaneous detonation. Nevertheless, as time progressed, less and less outside interference was needed to set the explosion off. Even the normal fluctuations of the let-down system could cause it to explode.

NRC had also asked consultants working under Deputy Director Robert Budnitz of the Office of Nuclear Regulatory Research to investigate the consequences of an explosion. He reported that an explosion would cause a pressure pulse of 5500 psi.[13] It could easily rupture and perhaps fragment the top of the core vessel. It would certainly blow out valves and seals and pumps needed to fight the explosion. As Budnitz explained to the commissioners, "[t]here was a time only yesterday when people were saying that, well, if it burns, it burns."

To the group at Bethesda, an explosion at this point seemed to almost insure the permanent loss of core cooling. Though no one dared say the words aloud, a meltdown was beginning to look almost probable. With consequences this severe, and with Denton apparently unwilling to accept that there was such a danger, Hendrie and Mattson decided that they must fly to Three Mile Island to personally warn Denton before President Carter arrived.

Denton was at the National Guard facility in Middletown, where the President's helicopter was expected to land. Hendrie and Mattson caught up with him literally minutes before the President was scheduled to arrive. Hurriedly, Mattson began to explain his calculations and conclusions.

But Vic Stello was also there: he had also done the calculations, and he would have none of this.[14] He challenged Mattson's assumptions. Mattson began listing all the experts at Bethesda who agreed with him. The debate threatened to grow into a shouting match. Stello told Mattson that he was "crazy" if he believed those calculations. Stello could not prove his case, but he was sure Mattson was "nuts."

Harold Denton listened to the scene as it rapidly got out of control. Denton tended to agree with Stello's more optimistic view. But the experts and the formulas rather clearly supported Mattson. From all this, Denton would have to cull a report for the President.

But there was no time. The helicopter arrived, and the President and First Lady stepped out. Denton, who had been repeatedly encouraged on the phone by Carter to be honest, made a snap judgment on what he would say. When the President walked up, Denton introduced himself and briefly outlined both sides of the dispute.[15] Carter was already there, and the media was ready to see him tour the plant. Turning back now was out of the question and was never even mentioned. However, at least Carter now knew what he was walking into.

Just after 1:00 P.M., the entire motorcade traveled the last few miles to the plant. The Carters, Governor Thornburgh, Denton, and the requisite photographers all entered the north gate, put on rubber boots to avoid picking up contaminated particles from the floor, and started their now famous tour.

Vic Stello, continuing his earlier argument, took NRC Chairman Hendrie to the NRC trailer. There Stello convinced Hendrie that Stello deserved a little time to get some outside experts to help substantiate his claims on paper. Hendrie agreed that, for a while, the NRC would stay quiet and give Stello his chance.

While NRC politics were still brewing, President Carter was conducting a perfectly orchestrated media campaign to calm the fears of the nation. He was repeatedly photographed smiling as he toured the plant.

Next the Carters went back to Middletown, to the borough hall. There Carter delivered a meticulously worded public statement. He said that the reactor was "stable." However, he continued, the plant would need to make one shift to bring the core to a cold shutdown. Over the next few days, the Governor would decide whether there should be a precautionary evacuation during the shift. Carter asked the citizens to listen to the Governor, and he expressed his faith that the citizens would respond as needed.[16]

The statement was simple and even a bit condescending, but the effect was electric. For days, Middletown's residents had been tense and cooped up indoors due to the Governor's earlier advisory. Now the entire town was outside, and there were even some attempts at fatalistic humor.[17] There was the unmistakable message in Carter's statement that they still might have to evacuate. But the President had come to town. That was exciting in itself, and it indicated that the reactor must not be as dangerous as they had feared.

At least in Middletown it did not look that dangerous. In Bethesda, the more they looked at the figures, the worse things seemed to be. Hendrie was in Pennsylvania, and Gilinsky was at the White House making recommendations on the scenario for a possible evacuation.

Commissioners Bradford, Ahearne, and Kennedy were left with the realization that the bubble was becoming more sensitive with passing time. By midafternoon, they were getting to the point where a precautionary evacuation would extend into the night, causing additional difficulties. If they were going to recommend an evacuation, it needed to be soon.[18] By the next day, it might be too late.

They called Gilinsky at the White House. They wanted him to agree to a statement that unless Hendrie learned something new at the plant, the NRC was recommending a precautionary evacuation of two miles to be carried out immediately. Gilinsky agreed. Another NRC recommendation to evacuate was about to be issued to the Governor.

However, while Gilinsky was agreeing to Bethesda's statement, Hendrie was also calling the commissioners. Hendrie cut in to give them the news. Stello had found the mistake.

Stello had been frantically calling his own industry experts, including Bettis Laboratory and General Electric.[19] They both agreed with him that oxygen generation should be suppressed, but they could not identify the flaw in Bethesda's formulas. These were well-established formulas that were even officially endorsed by an NRC Regulatory Guide on measuring hydrogen and oxygen generation in the containment building.[20]

But then it hit Stello. This was no containment building. The formulas assumed the pressure to be at atmospheric levels, not at the actual 1000 psi. Everyone had been looking for a flaw in the formula or maybe an arithmetic error. The problem was that the formula was correct, but for a different environment. At 1000 psi, oxygen was not being generated at all. At no time could the reactor have exploded.

Stello told Denton, who immediately saw the "sudden light dawning."[21] Together they told Hendrie, But Hendrie was not as accomplished at the mechanics, so he wanted to hear how the industry people reacted. As more opinions flowed in, it was soon obvious that Stello was right.

Hendrie called Bethesda. Just as the commissioners were agreeing to call Hendrie to recommend an evacuation, he called to tell them that the whole thing had been a mistake. There would not be another ill-advised NRC recommendation to evacuate.

The crisis was far from over. But the complexion of the reactor was completely different than it had been before the oxygen miscalculation was discovered around 4:00 P.M. Sunday. From this point on, things began to go unexplainably well. Instead of having to retreat from earlier optimistic statements, the people at the plant began to retreat from earlier pessimism.

NRC Chairman Hendrie stayed behind in Pennsylvania when President Carter returned to Washington. He had played an integral role in buying Stello the time he needed to discover the mistake. Since Hendrie was still there, he accompanied Denton on his regular nightly briefing of the Governor in Harrisburg.

There was still much to talk about, but the big news was that the bubble had reduced itself in size from about 800 cubic feet earlier that day to about 300 cubic feet. There was no immediate explanation for this, but it was obviously good news.

To avoid admitting that the NRC had made such a fundamental error, Hendrie did not mention the oxygen generation scare. It was some time before Thornburgh discovered what had been going on in Bethesda.

But Thornburgh asked Hendrie a crippling question when he reminded Hendrie that he was supposed to find out why Collins had recommended an evacuation on Friday. Hendrie was surprised by the question, obviously did not want to discuss it, and apologized for his own statements about precautionary evacuations.[22]

It was late Sunday evening, and Thornburgh needed to make some decisions on school closings for the next day. Denton still could not say that the crisis was over. If hydrogen was leaving the core vessel, it had to be entering the containment vessel. That was the set of circumstances that the formulas had been prepared to address. An explosion in the containment building still seemed possible.

For days, Denton had been trying to get at least one hydrogen recombiner hooked up and operating. Without it, a new and different explosion crisis was inevitable.

Besides, the reactor was still being cooled under pressure by one reactor coolant pump. If the pressure or the pump encountered difficulties, even the smaller bubble could expand to uncover the core. It was no longer as likely, but it could happen.

Earlier in the evening, Thornburgh had announced that state offices would open on schedule Monday. Employees could take administrative leave time if they so chose. Pregnant women and mothers of preschool children within five miles could miss work without having to take official leave. For schools he "recommended" that those within five miles stay closed. More distant schools were left to their own discretion, and most finally opened. As mentioned in Chapter 12, without knowing it, Thornburgh had removed the school busses required to make an evacuation practical.

After the meeting with the Governor Sunday night, Denton ended his exhausting day and went to sleep. When he returned to the control room the next morning, the news was so favorable that it was baffling. It was nice to have improvements, but only if they made sense.

The bubble was down to 150 cubic feet, having shrunk to half its previous size during the night. That could indicate trouble, since the hydrogen had to go somewhere. It should be entering containment, where the recombiners could not handle the load.

But to the surprise of everyone, the percentage of hydrogen in the containment atmosphere was not rising.[23] Rather, as Denton discovered on Monday morning, he had been unjustifiably pessimistic about the condition of the reactor all along. In addition to the oxygen scare, it now appeared that the "hydrogen bubble" had been comprised primarily of steam. That steam was now escaping to containment or simply converting back to pressurized water. Radiation levels were rising in the containment building, indicating that something other than steam was being vented. But instead of explosive hydrogen, the bubble was losing nonexplosive xenon gas.

Denton was cautious. It was tempting to report progress in the reactor; the public needed some good news. But as Denton spent Monday morning watching his pessimistic view of the reactor crumbling, he realized that if he now went to the press with this new information, it would make NRC's past announcements look irresponsible. If his new announcement proved to be prematurely optimistic, his credibility would be lost forever. So he waited.

But Met Ed would wait no longer. Twice on Sunday they had issued internal updates to their staff on the shrinking bubble. As instructed, the staff kept these quiet. At 8:30 A.M. on Monday they finally proudly announced that the bubble was so small that it may have disintegrated.

For George Troffer, a Met Ed technical briefer at the Hershey press center, sitting on this news in the midst of a crisis made no sense. He leaked the news to a radio reporter.[24] Soon the press was clamoring for a confirmation from Denton. Denton's hand had been forced. He would now have to describe the good news while explaining why the dire warnings of yesterday had been given.

Denton talked to Mattson about the inevitable press conference. They agreed that until things inside the reactor were better understood, they would have to be cautious and "save some wiggle room in order to preserve credibility."[25] So at 11:15 A.M., when Denton addressed the press, he said the bubble had shown a dramatic decrease, but probably not as far as Met Ed was claiming.

Denton was asked about the explosion possibility. He said that the "emerging consensus of technical opinion" was that the actual evolution rate of oxygen was way below the "very, very conservative" estimates of the day before. It was perfectly worded, giving neither the press nor the Governor any hint that the NRC had created a panic over what could easily be called a dumb mistake. In fact, Denton waited for more than another day before announcing that explosion was no longer a consideration. By that time, the bubble was gone.

After Monday afternoon, the evidence that the crisis had ended was unmistakable in the control room. Denton talked to the commissioners that night to get permission to not even try to depressurize. With the bubble gone and the core obviously covered and the primary loop almost watertight, there was no obvious advantage to going to residual heat. Instead, the plant could spend its time working on cleaning up the remaining points of radioactive releases.

There were still a couple of battles to be fought in the political arena. Robert Adamcik, the federal on-site coordinator, had a few problems convincing federal agencies that they had to respond to him, even if there was no declaration of emergency. He finally called the White House over the weekend to verify his authority.[26] After that, when he met resistance, he reminded the other agencies of that call. Beginning Sunday, he had daily coordinating meetings in Harrisburg.

The more dramatic battle involved the potassium iodide. The first shipment of 11,100 bottles that arrived at Harrisburg was in terrible condition, with well over half the bottles having no labels, obvious contamination, or visible leakage.[27] Later shipments improved considerably.

When Pennsylvania Secretary of Health Gordon MacLeod received them, he stored them at a warehouse in Harrisburg. But when Jack Watson discovered on Sunday that the state evacuation planners did not know what the federal agencies were recommending, he asked HEW to recommend a dissemination plan to the state.

On Monday, Director Donald Frederickson of the National Institute of Health recommended to John Villforth at FDA that the drug be administered immediately to plant workers and be available on one hour's notice to those living within twenty miles.[28] Frederickson's source of information was a Sunday NRC report when the bubble was still threatening and iodine was known to be in containment.[29] It was Tuesday, April 3, before the recommendation reached Jack Watson with HEW approval. By Tuesday, the recommendation was ridiculous, but it was forwarded to the Governor.

Thornburgh had a meeting at 1:00 P.M. Tuesday with his health people and federal representatives to try to understand the unexpected recommendation. Even though there were small off-site releases of iodine detected in some milk on Monday, Secretary of Health MacLeod argued that the recommendations were now completely out of line. The radiation was too low; the side effects were small but did exist; the quality control on the bottles was low. Finally, giving the unfamiliar vials to citizens and explaining their use would start the panic all over again. Thornburgh was convinced, and the idea was dropped.

But Joseph Califano testified before a Senate subcommittee the next morning and made the amazing statement that the doses were still needed, but that the state refused to administer them. Califano said that he intended to call the state later that day to try to convince them again.

Califano's statement was irresponsibly uninformed and was tantamount to an accusation that the state was denying its citizens a needed anti-cancer drug. Still, the various health officials in HEW were baffled when they found out that Secretary MacLeod was furious. At least two of them called MacLeod to explain how supportive they felt they were being.[30]

Fortunately, little serious cooperation was still needed between the state and HEW. The potassium iodide bottles did not leave the warehouse until FDA finally reclaimed them and sent them to Little Rock, Arkansas, for storage.[31]

The NRC has a traditional definition of a cold shutdown in a reactor—when water at atmospheric pressure covering the core will not boil. For several days, Thornburgh had asked Denton if the advisory for preschool children and pregnant women could be raised. For days, Denton stalled by asking Thornburgh to wait for depressurization and a cold shutdown.

But by Monday evening there was tremendous and growing resistance in the control room to any plans to shift the plant's status. The cooling loop was obviously fixing itself. Therefore, by Tuesday morning the idea of a planned depressurization was all but abandoned.

Eventually, the Hershey evacuation center closed for lack of people. Eventually, the citizens went home. Eventually, the Governor raised the advisory. Eventually, the hydrogen recombiners were hooked up.

The crisis at Three Mile Island never had a date on which it ended. It just faded away.

NOTES

1. Deposition of Jessica Tuchman Mathews for President's Commission on the Accident at Three Mile Island, Washington, D.C., August 23, 1979, pp. 117–18.

2. Ibid., pp. 118–20.

3. Deposition of Roger Mattson for President's Commission, Bethesda, Maryland, August 6, 1979, p. 194.

4. Ibid., p. 194.

5. Testimony of Harold Denton, President's Commission, May 31, 1979, pp. 310–11.

6. "Report of the Office of Chief Counsel on Emergency Response" to the President's Commission, October 1979, p. 126.

7. Deposition of Roger Mattson, p. 193.

8. Deposition of Jessica Tuchman Mathews, pp. 118–20.

9. Ibid., p. 121.

10. "Report of the Office of Chief Counsel on Emergency Response," p. 130.

11. Deposition of Jessica Tuchman Mathews, p. 122.

12. Ibid., pp. 125–30.

13. NRC meeting transcript, April 1, 1979, section 2, p. 27.

14. Deposition of Roger Mattson, p. 193.

15. "Report of the Office of Chief Counsel on Emergency Response," p. 126.

16. Jimmy Carter, press statement, Middletown, Pennsylvania, April 1, 1979.

17. "Report of the Office of Chief Counsel on Emergency Response," pp. 126–27.

18. NRC meeting transcript, April 1, 1979, section 3, pp. 5–7.

19. "Report of the Chief Counsel on Emergency Response," p. 131.

20. Deposition of Harold Denton for President's Commission, Bethesda, Maryland, August 2, 1979, p. 105.

21. Ibid.

22. "Report of the Office of Chief Counsel on Emergency Response," p. 134.

23. Testimony of Charles Gallina, President's Commission, May 31, 1979, p. 294.

24. "Report of the Public's Right to Information Task Force" to President's Commission, October 1979, p. 221.

25. Ibid., p. 221.

26. Deposition of Robert Adamcik for President's Commission, Washington, D.C., August 8, 1979, pp. 58 and 77.

27. Deposition of Gordon MacLeod for President's Commission, Harrisburg, Pennsylvania, July 23, 1979, pp. 71–72.

28. Deposition of Donald Frederickson for President's Commission, Washington, D.C., August 2, 1979, p. 47.

29. Ibid., pp. 55–56.

30. Deposition of Donald Frederickson, p. 64.

31. Deposition of Tom Gerusky for President's Commission, Harrisburg, Pennsylvania, July 24, 1979, pp. 93–94.

Chapter 14

Prologue or Epilogue?

Technology has been a seductive force in American life.
For almost one hundred years, it has dazzled us with almost continuous improvements in our standard of living.
It has replaced the horse and railroad with automobiles and supersonic aircraft. Instead of telegraph wires, we have telephones, televisions, and satellite microwave communications. The kindly but
ill-equipped family doctor now offers antibiotics, organ transplants,
and microsurgery. Clerks have been replaced by computer program-
mers and typists with machines that print at several lines per second.
For leisure, technology has given us computer video games and disco
strobe lights. It satisfies every mood from stereophonic classical
music to Doris Day movies to football instant replays.

We have not achieved all this without a cost. Society is very different in the technological age, and not all of the changes are uni-
versally welcomed. But part of the difficulty in identifying the costs
of technology is that almost every aspect of technology contains
both costs and benefits.

Many argue that technology has destroyed the social fabric of the
American family. To some extent, the soaring divorce rate is an
undeniable consequence of the new era. But technology has also
freed many from adolescent mistakes that once were glued together
by economic necessity. It has eliminated the requirement to raise
children as a form of social security, introducing other avenues of
emotional fulfillment to both sexes. Finally, it has offered greater
medical and financial assurances that solid emotional relationships
are less likely to be torn asunder by the whims of nature.

Technology has undeniably created air and water pollution. But it has also all but eliminated the smells of horse and human excrement from the streets. Coal soot no longer turns the snow black. The noxious by-products of life have new additions, but are finally somewhat controllable, depending on how much of technology we divert to fighting the effects of industrial pollution, natural silting, pestilence, and drought.

But there is one cost of technology that may not have an adequate compensation in the long run. Like many of the costs, it does not particularly relate to the physical environment; this cost is a change in the national psyche. Technology has required from us a new societal faith.

Our technological contraptions have become too complex for most of us to fix or to understand. Most components of the technological services we receive are even removed from our sight. If we are to enjoy the benefits of their existence, we must believe without prior knowledge that these tools of technology are compatible with our future best interests.[1]

There are two tenets to this twentieth century faith in technology. First, we must believe that any technological device that has the potential to be truly threatening to our way of life has a companion mechanism that can control it. We must also believe that there are people who understand the control mechanisms and who are making sure that the controls are in place.[2]

If we can only believe in our society's technology, the benefits are as great as in almost any religion and more visible than in most. There are the material comforts of mobility, health, and climate control. But there is also the inner peace that comes from a sense of invulnerability. While cars and planes may crash and misapplied medicine may cause occasional deaths, the societal aura of inner peace continues so long as life's duration and comforts continue to grow. We can accept thousands of automobile deaths per month because the option is to do without the benefits and the protection provided by cars.

The faith has always had its heretics.[3] But they have normally been so few that, like the Amish and Mennonite families downstream from Three Mile Island, they have been more cute and quaint than threatening. The faith has been far more subject to revisionists. It is perfectly acceptable to argue that automobile emissions need a technological "fix." One can even argue that one particular contraption, such as the hydrogen bomb, does not offer the peace and fulfillment of the rest of the tools of the faith. If society accepts that, we can moderate our enthusiasm with test ban treaties or other plans for political control.[4] We do not offer to abandon the technology that made all of this possible.[5]

To each of us, whatever faith we profess offers some benefit or benefits that make the faith subject to an ultimate test. The benefit may be nothing more tangible than peace of mind, but if that peace of mind ever becomes untenable, the faith of the individual eventually crumbles.

For technology, a fundamental benefit of the faith has been the sense of invulnerability. We have accepted that some of our technological devices have the potential to be catastrophically mishandled. But whatever stupid human errors or technical malfunctions might occur, we have not seriously come to grips with the thought that anything with the potential to destroy our societal progress could ever totally get out of hand.

We may have computer simulations that predict the virtual collapse of the industrial world within one hundred years.[6] But we cannot bring ourselves to abandon the technological life described in those dire predictions. Instead, it is the second tenet of our faith that the guardians of the technology will not allow such a catastrophe to happen.

It is that second tenet that has been called increasingly into question since the accident at Three Mile Island. For many inside and outside the nuclear energy industry, the events of March and April 1979 reopened the question of whether the guardians of our faith have been doing their job. The answer that we as a society finally accept to that question determines much of the future of nuclear energy in this country.

The evidence that society has reopened this question is all around us. Within weeks of the accident, the NRC offered a technological "fix" of the accident through NUREG 0560 and NUREG 0600.[7] Both were staff studies generated by experts with top credentials changing the geometry and operating procedures to make the chain of events at Three Mile Island impossible to duplicate in the industry. There was no such guarantee offered after the Brown's Ferry nuclear plant fire, but Brown's Ferry did not challenge our faith in technology. There were no off-site releases or public awareness that there was such a danger. At Three Mile Island, these violations of our sense of invulnerability occurred, and the public concern continues beyond the technological "fix" of that particular accident.

There have been other offers of a technological "fix" of the accident. The NRC's Division of Inspection and Enforcement imposed a moratorium on new reactor license applications while it assessed the lessons of Three Mile Island. But the NRC's "Lessons Learned Task Force" reported suggested changes in procedures in July 1979. On August 22, 1979, having implemented those changes, Harold Denton reopened the processing of applications.[8]

There was an immediate outcry in Congress and in the press. The President's Commission immediately subpoenaed Denton, and he was called to testify the very next day. As contrasted to Denton's earlier appearance, when he was treated with the respect appropriate to his reputation, Denton was now approached with outrage. For hours, the commission members verbally accosted him for what they considered to be an inexplicable, inexcusable error.[9] The President's Commission finally voted to go over Denton's head and to speak to the NRC commissioners directly. The commissioners finally reimposed the moratorium.

To the President's Commission, the "fix" was in the administrative machinery that inspects reactors and does the emergency planning. They specifically warned that "fundamental changes must occur in organizations, procedures, and, above all, in the attitudes of people. No amount of technical 'fixes' will cure this underlying problem."[10]

The President's Commission proposed changes in the form of a new structure headed by a single administrator who clearly managed the agency. The new agency was to be strongly checked by two independent safety committees. The agency was to have a more clearly defined obligation to protect the safety of the plants and, therefore, the population.[11] In a similar vein, the commission suggested changes in the process of operating reactors and in state emergency plans.

Other proposed "fixes" have run the gamut of possible actions. The political far left has provided the most interesting contrast. When the leftist factions disagree, they can border on the comical. One faction insists that the whole accident was the result of capitalist sabotage and that the proper "fix" is to immediately build 1000 more plants to liberate the Third World. Other factions want existing plants dismantled and converted to fossil fuel plants or parks or whatever.

Sitting somewhere on this spectrum is an American public that has more questions than answers. One question—what actually happened at Three Mile Island—has hopefully already been addressed with sufficient detail and breadth to satisfy most needs. But the more important question of why Three Mile Island happened leads to answers that have more of an impact on our faith in technology than on the future of nuclear energy. Furthermore, the explanation lies more in the processes of high technology than in the specifics of the Three Mile Island crisis.

We want to view nuclear energy, like all potentially dangerous high technology, as proactive. We want to believe that someone is there before the accident, checking the parts and analyzing the diagrams to make sure that the chance of something going wrong is as small as is reasonably possible. It is part of our faith that regulation by experts with integrity is continual and reasonably effective. Regulation does not have to be done by "regulatory agencies." In fact, in a time when government is held in low esteem, it is common to argue that the manufacturer can do it better. It is only important that someone is assuring that technology is restricted to activities that serve the public interest.

If we are to venture a guess as to whether the regulatory process in high technology is working, we must examine it across the three phases where dangerous technology has the potential to become catastrophic technology. First, there is the design engineering phase. Often before the first shovel of dirt is moved or the first piece of metal is cut, judgments are made on the potential risks of a proposed new device. Safety systems are then designed to hopefully counteract these risks. If the design engineering phase does not include adequate safety provisions, even heroic actions later may not be sufficient to prevent catastrophic failures.

Second, some sense of regulation must exist during the production phase. Safety systems are useful only if they are produced to adequate levels of quality assurance. As we learned at Three Mile Island, safety components that do not operate properly can complicate or even cause accidents.

Finally, the operators of the high technology must be adequately trained in emergency procedures. It is not sufficient to train them in the subtleties of a properly operating machine. They must also be properly schooled and drilled in the process of fighting severe malfunctions.

It is against these standards that the experience of nuclear energy and other high technologies must be judged. At first look, the regulatory process in high technology seems to be doing well. Manufacturers and government agencies are cluttered with what appears to be constructive activity. New airplane designs are rigorously tested in wind tunnels and computer simulations. The FAA then conducts tens of thousands of inspections before certifying a plane as airworthy. The Food and Drug Administration can be painfully slow and demanding in certifying a new drug as beneficial and safe. The NRC licenses and then inspects each individual reactor.

In breaking the regulatory process into its constituent parts, however, serious gaps are immediately evident. Like many of our activities, there is often more symbol than substance in the regulation of high technology.

The first phase of high technology discussed earlier is design engineering. For nuclear reactors, there are two aspects of the regulation of design engineering that do not correspond to the assumptions stated earlier as part of our societal faith in technology.

First, there is confusion over who is responsible for the regulation. In most of the high technologies, there are government agencies that literally have the power of veto over new products. For nuclear reactors, that agency is the U.S. Nuclear Regulatory Commission. It makes sense to us as the consuming public that regulation should be done by a government agency without a vested interest in the product. Otherwise, manufacturers are forced to decide how many safety features will be added to the final sale price of their technological items. However, the American public is not necessarily aware of how many of the decisions are actually made by the manufacturer.

An example from aeronautics provides a useful illustration of the problem. On the DC-10, one of the less celebrated safety achievements of commercial aviation, there were 42,950 inspections by the Federal Aviation Agency to certify that the design met the spirit and the law of airworthiness standards. But the FAA had enough personnel to conduct only 11,055 of the inspections.

The other 31,895 inspections were conducted by "designated engineering representatives"—employees of McDonnell Douglas Aircraft Corporation who also spent part of their time certifying DC-10 component designs as airworthy.[12] The spirit of potential conflict of interest was most notably strained on those occasions when design engineers were called upon to certify their own work as meeting FAA airworthiness standards.

There is nothing particularly insidious about this arrangement on the part of McDonnell Douglas or the FAA. It is common throughout the industry, and in some ways it is a more rigorous inspection system than that used to certify nuclear reactors. In both industries, the regulating government agencies have only a fraction of the personnel that they would need to do the job themselves.

Regulating the design phase in nuclear reactors is more complicated than for new aircraft. To begin, there are three periods during the construction of a reactor in which the NRC has heavy regulatory input into the design of safety systems. The first begins months or even years before any formal applications are filed and consists of informal discussions between the eventual applicant and the NRC staff. These discussions assure that the NRC is not adamantly opposed to the general plans for the reactor, including the site chosen or the contractors to be used. This information is vital for the utility company before it invests large sums of money in land acquisition and designs for a reactor that stands no chance of being approved.[13] However, since this phase occurs before those funds are spent on sophisticated design studies, safety systems are normally not yet described in detail and cannot be discussed at much length.

The first detailed design proposals are submitted during the second phase, during which the utility company files an application with the NRC for a construction permit. Along with the application, the utility company must file a preliminary safety analysis report (PSAR). The PSAR contains, among several other sections, "a 'preliminary' analysis of systems to be provided for prevention and mitigation of accidents" and "a showing that the safety issues involved can be resolved before construction is completed."[14] The wording of the NRC in describing the PSAR emphasizes the degree to which it is preliminary and can be very nonspecific. In fact, since the PSAR is available for public review and can be used against the application during the required public hearings, there are strong incentives for the company to keep the PSAR vague.

Review inside the NRC during the application for a construction permit is nominally assigned to the three member Atomic Safety and Licensing Board that holds the public hearings. In fact, a number of offices and staff personnel inside NRC study the proposal and report back to the board. The most conspicuous of these is the Advisory Committee on Reactor Safeguards (ACRS), which is an advisory group consisting of fifteen part-time nuclear experts and a staff of twenty.[15] While the ACRS has no power of veto, it studies the PSAR and renders a professional judgment to the NRC.

However, the ACRS and the NRC normally bypass serious attacks on proposed safety designs until the third phase, after most of the construction is completed. For instance, when the ACRS reviewed the construction permit application for TMI Unit II, they intentionally bypassed two design problems that later contributed to the severity of the accident. According to their letter of report to the NRC:

> The applicant has been considering a purge system to cope with potential hydrogen buildup from various sources in the unlikely event of a loss-of-coolant accident. . . . The Committee believes that this matter can be re-

solved during construction of the reactor. . . . The Committee recommends that a study be made of the possible consequences of hypothesized failures of protective systems during anticipated transients, and of steps to be taken if needed. The Committee believes that this matter can be resolved during construction of the reactor.[16]

There are a couple of reasons why the analysis of safety systems is delayed until the last possible stage of licensing. First, nuclear energy is a rapidly developing technology that requires a long construction time. It is the belief of the NRC, and more specifically of the ACRS, that many technical barriers evident during the design phase of a reactor can be "worked out" by advances in nuclear engineering as construction continues.[17]

There is considerable evidence to support the contention that safety engineering in a reactor progresses during construction. In fact, the NRC has even structured the review process around this assumption. Those safety questions left unanswered in the PSAR filed with the construction permit application are normally delayed until the completely separate Final Safety Analysis Report (FSAR), filed with the application for an operating license.

One characteristic of the process of delaying safety design systems until the last possible phase is that no two reactors are exactly alike. When the FAA or its designated representatives certify an airplane design as airworthy, several copies of that design are produced without needing new certification. When changes are finally made because of the lessons of experience, the FAA can concentrate on the changes. In reactors, where different contractors make continual changes in the design of the nuclear steam system, the nonnuclear system, the control room, and so forth, every reactor becomes a virtual prototype. Each must be studied in considerable detail, and more time is available for this study if the difficult issues are delayed until the final stage of licensing.

The final stage is the application for an operating license once the reactor is almost completely built. This is also the stage during which the FSAR is filed as the final description of the safety systems to be used in the reactor. For reasons outlined earlier, this is the stage during which the more difficult safety problems are addressed. Unfortunately, two problems commonly arise in addressing serious safety questions at this stage of the licensing process.

First, the operating license is requested after more than a billion dollars have been invested in the construction of the plant. The delays granted for Three Mile Island's PSAR suggested that the remaining problems would be solved during construction, not after construction. Once the FSAR is filed, the proposed solutions have normally already been built.

The NRC has the right to reject the operating license because of unsolved or inadequately solved safety problems. In the jargon of the industry, this would require that the designs be "ratcheted" or that an already constructed component be "backfitted." Occasionally, relatively expensive changes or additional studies are required before the operating license is granted. However, both the NRC and the utility companies agree that it would be unreasonable to require extensive changes once the reactor has progressed this far.[18] As a result, no reactor has ever been abandoned at this stage, and no operating license has ever been denied.[19]

Even the NRC does not agree that all safety problems are solved by the time a reactor is allowed to go commercial. But since the NRC is not willing to abandon nuclear energy because of unsolved safety problems, the agency has developed two mechanisms for allowing problems to be ignored during the licensing for operation stage.

First, they can classify particular problems as "generic." Any problem that exists in all reactors or in all reactors of a particular design are considered generic and cannot be used to stop an application for a particular reactor. Instead, consideration of the generic problem is delayed until a solution is found for all reactors that are affected. There is some disagreement on how many generic problems remain unsolved by the NRC. Robert Pollard, formerly a design reviewer for the ACRS, has testified that when he resigned in 1976, there were more than 200 unsolved generic issues on reactors in the United States.[20] In January 1978 the NRC produced a list of 133 generic issues to answer a congressional order. However, Robert Pollard claims that the list was as large as ever, but had been reduced by combining similar generic problems. In the NRC's 1979 report to the Congress, the list was further reduced to seventeen "highest priority" items. Whatever the actual number, there are unresolved safety problems in all reactors that cannot be used to stop the application of any reactor because the problems are termed "generic."

On the average reactor there are about thirteen to fifteen safety-related unresolved questions that reach the operating licensing stage but that are clearly not generic.[21] Instead, they are problems created when the particular contractors combine their products given the current level of technology. They create problems that relate to individual reactors. If the NRC is to stop reactors at all once they reach the operating licensing stage, these are the most logical safety problems to use as a justification.

Instead, the NRC regularly issues the license, but uses an administrative remedy that amounts to a fourth stage of licensing. A reactor that is still under construction, which means that it does not yet have an operating license, is administered within the NRC by the Division of Project Management. Once the reactor is licensed to operate, it is supposed to be shifted within the NRC to the Division of Operating Reactors. In fact, the Division of Operating Reactors refuses to accept the responsibility for reactors that have open safety questions. As a result, new reactors often spend months or as long as a year in a state of limbo between two monitoring NRC divisions, neither of which accepts full responsibility for the reactor.[22]

On February 8, 1978, TMI Unit II was granted an operating license but was left under the control of the Division of Project Management because of unresolved safety issues. In September, when the Division of Project Management tried to transfer the reactor to the other division, the Division of Operating Reactors refused since there were still fourteen unresolved safety issues.[23] When the reactor had its accident on March 28, 1979, it still was not recognized by the NRC as an "operating reactor."

The second aspect of technology that must be regulated if the technological process is to operate in the public interest is the production phase. Equipment must be maintained in good operating condition or the reactor will not respond as was intended during the design engineering phase. The NRC maintains an inspection force in its Division of Inspection and Enforcement, which has personnel in each of the regional offices. However, the NRC has never pretended that it has enough inspectors to collect usable information on the operating condition of reactors.

Instead, each licensee is required to report to the NRC regional office any "event or unusual occurrence" that happens at the plant. The utility companies appear to fulfill this requirement with some degree of good faith, and over 3000 reports are filed with the NRC annually. The NRC inspectors then use the information gathered through the reports to decide if there are any patterns or lessons that should be passed along to other reactor licensees.

However, to save time and effort, the NRC restricts its regulation and most of its attention to what it defines as "safety-related systems."[24] A safety-related system is one that is essential to the safe operation of the plant during an emergency. However, the NRC's definition of what constitutes a safety-related system is more restrictive than experience would indicate is appropriate.

A code safety valve, which releases serious overpressures from the pressurizer, is a safety-related system since its malfunction could lead to an explosion in the nuclear steam system. The pressure relief valve that stuck open at Three Mile Island, however, is not a safety-related system. Should the valve stick closed, the NRC reasons, the code safety valves could relieve the pressure. Should it stick open, there is a block valve downstream to stop the unintended release. Neither the pressure relief valve nor the block valve is safety related, the NRC maintains, since the existence of each assures that the proper operation of the other is not necessary for the safety of the plant.[25]

On March 29, 1978, the day after TMI's Unit II went critical, the pressure relief valve failed in an open position when it lost electrical power. The regional inspector for the NRC questioned the wisdom of designing the valve so that it opened when it lost power. The utility answered that the wiring of the valve met the requirements of the FSAR since the valve was not a safety-related system.[26] The NRC has no specific definition of "safety-related," and different NRC personnel understand the term to include different components.[27] The GPU corporation had the authority to put the pressure relief valve into the nonsafety category unless the NRC objected. the NRC did not object, and the valve was still largely ignored despite its poor operational performance until the accident on March 28, 1979.

Since the NRC cannot be available to witness and analyze all operating experience, the agency has developed one shortened maneuver that has a continuing impact on safety maintenance. Each reactor is required to get NRC approval on all its emergency and maintenance procedures. For each possible malfunction and for each testing procedure, the company is required to have a written set of steps to be followed and forms to be checked as the steps are completed.

However, the NRC has some difficulty in using these procedures to gain much control over the reactor. When NRC licensed Unit II at TMI, they had enough personnel to commit only twenty hours by one inspector to all the procedures for emergencies and normal maintenance. The inspector found no flaws in the procedures.[28] However, changes made in these procedures after initial licensing were never rechecked. These included the decision in August 1978 to allow emergency feedwater valves 12A and 12B to be closed during testing while the reactor was still at full power. This procedure clearly violated an NRC Technical Specification, but was not detected by the NRC.[29] When the accident started on March 28, 1979, two days after those valves were closed for normal testing, they were found once again to be closed.

Another difficulty is that the procedural manuals at best can assure that the utility discovers that some systems are inoperative or have difficulty operating; they cannot assure that the utility does anything about the malfunctions. For instance, Met Ed knew that its pressure relief valve had a history of failing in an open position. Since the reactor opened, it had continually leaked and registered in the 175 to 195°F range. This author has personally visited a different reactor that was operating at full power with the pressure relief valve annunciator turned off since leakage was in excess of 200°F. At Three Mile Island, James Floyd was worried during the crisis on Friday because he knew that the pressure relief valve on the makeup tank had a poor operational history. Met Ed had been complaining to GPU since before the reactor was built that the condensate polisher system was inadequate. When the accident started, makeup pump 1A, which was a part of a safety-related system, was known to be breaking down. Met Ed had not investigated far enough to discover that the problem was a relatively simple malfunctioning switch.

Finally, the procedural manual approach to an accident assumes that the operators in the control room recognize the nature of the accident and that they use the correct procedures manual. Given the general nature of the manuals, so that one is written for each major type of transient or accident that might occur, this has not seemed to the NRC to be a significant assumption. However, it was over two hours into the TMI accident before the operators realized that they had a stuck pressure relief valve, requiring the manual for a "small break loss of coolant accident," and not a simple turbine and reactor trip.

This last difficulty leads into the third area of reactor operations in which regulation is necessary in order to protect the public safety. It is an inherent part of the faith of technology that people are monitoring the technological machinery to ensure that no extraordinary chances are being taken to endanger the public. We insist as a partially informed public that the final monitoring be left to informed human experts. However, that part of the faith assumes that the humans left in control are properly trained to handle emergencies. There is considerable evidence that reactor operators may not be trained in emergency procedures to the extent assumed by the public.

One task force within the President's Commission identified five operator errors during the early hours of the TMI accident that they attributed to a lack of training.[30] First, the operators had been incorrectly trained that the volume of coolant in the primary loop was always reflected in the pressurizer level. During the accident, the pressurizer level rose as the volume of water in the primary loop dropped. Second, the operators failed to recognize that the pressure relief valve had stuck open. Third, the reactor reached saturation conditions so that steam was forming throughout the coolant loop, but the operators failed to recognize the fact. Fourth, the operators showed little knowledge of the thermodynamics of nuclear cores. Specifically, when the open pressure relief valve was blocked, it was logical to expect the core to heat since a source of heat release had been removed. The core did begin to heat, but the operators did not understand why and decided that the core thermocouples must be malfunctioning. Also, the operators apparently did not understand that if both the primary and secondary loops are at about 1000 psi, almost no heat will transfer between the loops. Fifth, when the radiation levels began to rise as early as nineteen minutes into the accident, the operators failed to interpret this as evidence that the core was overheating.

If the operators committed this many mistakes or even a substantial portion of them, it introduces some serious questions about the qualifications and training required for reactor operators. Surprisingly, the NRC has almost no qualification or training requirements for operators and relies exclusively on oral and written tests of personnel selected by the utility companies. The NRC does not require even a high school level of education. It does not require that the candidate be in good physical or emotional health. If the candidate can somehow pass the examinations, he need not have any training or experience in nuclear energy. The NRC offers no courses. Instead, all training and screening of candidates is delegated to the utility company, with the exception that the NRC administers a final oral and written examination before the candidates are licensed as operators.

There is also no requirement that operators licensed by the NRC have access to a simulator, and they cannot practice on an actual reactor until they are licensed. Instead, they are tested by a series of questions requiring that diagrams be drawn or that "what if" scenarios be talked through. The problem with this is that the operators can learn about emergency procedures only through hypothetical scenarios described in procedural manuals. They cannot experience the tension, confusion, or even panic conditions that might occur when an accident is actually in progress.

Because of this limitation, most utility companies have arranged for their trainees and operating personnel to practice accidents periodically on a simulator. However, they encounter the problem that control rooms and nuclear steam systems are built by different contractors. The industry long ago learned that there is more to be gained by simulating the nuclear steam system than by simulating all the various control room configurations offered by each contractor. It is easier for TMI control room operators to learn how Babcock and Wilcox steam systems work and transfer their training to a Burns and Roe control room than to learn the preferences of Burns and Roe on where controls should be located in the room without knowing how the systems operate.

Unfortunately, this means that any similarity between the simulator control room and the actual control room in which the operator will work is almost accidental. While this is the most practical approach to simulators that the industry has been able to produce, its limitations for emergency training are best illustrated by one of the coincidences of Three Mile Island. The accident could not have occurred in Unit I because the polisher bypass valve would have automatically engaged when the polisher became isolated. If this problem is ignored, and one assumes that a transient began in Unit I identical to the one in Unit II, several quench tank annunciators would have immediately told the operators that the pressure relief valve was stuck open. In Unit I and the Babcock and Wilcox simulator, these annunciators are in plain sight. In Unit II, the operators had to leave the center of activity and walk about forty feet to see the annunciators for the quench tank. Despite this crucial difference, which meant the difference between a simple transient and the accident at Three Mile Island, the operators for both control rooms were trained on the same Babcock and Wilcox simulator.

Simulators are just one of the tools used by Babcock and Wilcox to train operators, and the simulator is used to varying degrees over time as Babcock and Wilcox restructures its training courses. One of the more popular courses, which has been attended by fifty-four Met Ed employees since 1973, includes twenty-six hours of lectures, ten hours of assigned study, forty hours on the simulator, and a four hour written exam.

One possibly disturbing aspect of the course is that no one has ever failed. In fact, in all the training ever offered to Met Ed employees by Babcock and Wilcox, only two people have ever failed to receive whatever certificate was being offered for the course.[31] While high grades can be reflective of careful selection by Met Ed or excellent instruction by Babcock and Wilcox, there are a couple of indications that the high grades may be more reflective of a lack of rigor. For instance, in those courses where grading is done at all, it is usually in the form of written comments that the staff of the President's Commission found to be virtually meaningless.[32]

More disturbing for the safety of nuclear energy, while operators spend considerable time practicing system shifts and reactor startups, they spend amazingly little time practicing emergency procedures. Not everyone who took the course described above ever simulated a drill as simple as a reactor trip.[33] Most but not all simulated leaks in the primary and secondary coolant loops. Almost no one simulated a stuck control rod. In the six year history of the training program, no one ever simulated a stuck pressure relief valve. While the simulator is the only place an accident can be practiced, the simulator is actually used overwhelmingly to mimic what the operators do in the control room during normal operations.

But the most dramatic failing of the training program by far is that there is no guarantee that the information being taught is correct. The most relevant illustration of this in the case of Three Mile Island is the assumption in the industry that the pressurizer level reliably reflects the level of coolant in the core. On a number of occasions before Three Mile Island, in accidents that were often freakishly similar to Three Mile Island, that assumption was proven incorrect.

The first hint of difficulty happened at Westinghouse. Westinghouse does not use the two stage pressure relief valve and code safety valve combination. Instead, two relief valves open in unison during overpressurization. Westinghouse also takes the assumption that pressurizer level indicates the volume of coolant one step further than Babcock and Wilcox. Until the accident at Three Mile Island, the ECCS would not automatically engage in Westinghouse reactors until the primary pressure was low and the pressurizer level was low.

The fallacy in this logic became clear in August 1974. At a Westinghouse plant outside Beznau, Switzerland, a turbine trip led to overpressurization. Both pressure relief valves opened, and the reactor tripped. Without the knowledge of the operators, one of the pressure relief valves failed to reclose. Within minutes, the Beznau plant was having a Three Mile Island type transient. Primary pressure began dropping dangerously low, while the hole at the top of the pressurizer caused it to fill. Within about ten minutes, steam voids were forming throughout the primary loop, and the pumps began to cavitate. The operators did not turn off the ECCS because it never came on. It did not engage unless both the pressure and the pressurizer level were low.[34]

Twelve and one-half minutes into the transient, the operators figured out what must be wrong and saved Beznau from being the Three Mile Island accident. They manually engaged the ECCS, and they blocked the pressure relief valves. Within a few minutes, the plant was stable.

Because the plant was not in the United States, Westinghouse did not have to report the incident to the NRC. They did not report it until 1979. Also, even though the accident had shown that the pressure and the pressurizer level did not have to agree, Westinghouse continued for five years to build plants where emergency core cooling would not engage until the two did agree.

In the intervening years, Tennessee Valley Authority engineer Carlyle (Carl) Michelson became interested in the problem from an entirely different angle. Michelson was a consultant to the ACRS, and he backed into this issue through a related concern that the NRC always assumed that a safety system that could handle a large accident would also be appropriate for a small accident of the same type. Specifically, he did not believe that actions appropriate to fight a large loss of coolant accident were also appropriate to fight a small loss of coolant accident.[35]

During 1977, Michelson wrote a handwritten description of the problem he envisioned, using a Combustion Engineering reactor design as a model. He described the way the pressurizer level could remain deceptively high should one of the pressurizer valves stick open. He passed his document around the ACRS, but no action was taken because of it.

On September 1, 1977, Michelson produced a typed copy, but this time he described the way the scenario would work in a Babcock and Wilcox reactor. The second report was almost an exact description of what finally happened at Three Mile Island. Furthermore, the report speculated that the operators would be fooled into turning off the ECCS because Babcock and Wilcox had trained them to watch the pressurizer level.[36] Michelson also began circulating this memo within the ACRS.

If there was any question of the credibility of Michelson's scenario, that was resolved on September 24, 1977, when his accident actually happened. Toledo Edison's Davis–Besse Unit I in Ohio was a Babcock and Wilcox plant virtually identical to the nuclear steam system at TMI. It had just been refueled and was being restarted when the turbine tripped and then the reactor tripped. The pressure relief valve stuck open, although the indicator said that it had closed. Within minutes, the primary pressure began to fall dramatically while the pressurizer level climbed. When the pressure fell below 1600 psi, the ECCS engaged. The operators discussed the contradiction between the pressure reading and the pressurizer level. Because Babcock and Wilcox training sessions had emphasized the importance of the pressurizer level and had repeatedly warned that the pressurizer should never be run solid, the operators finally manually overrode ECCS and cut the water flow to the core. Within minutes, voids were forming in the hot legs and the top of the core vessel.

The sequence of events at the Davis-Besse reactor was identical to what happened at TMI, except for two important differences that saved the plant. First, the mistake was discovered and corrected after twenty minutes at Davis-Besse, not after two hours and twenty minutes. More important, Davis-Besse entered the transient at only about 9 percent power. Since they had just refueled, the core had almost no residual heat. In identical circumstances, the core at Davis-Besse was simply not as sensitive as it was at Three Mile Island or as it would have been at Davis-Besse had it contained old fuel at full power.

Babcock and Wilcox sent an inspector, Joseph Kelly, to the plant to investigate the accident. After a couple of days he returned to Lynchburg and reported what had happened to a gathering of about thirty Babcock and Wilcox employees. Most of the engineers listening were interested, but not overly concerned. However, Bert Dunn was in charge of the ECCS unit at Babcock and Wilcox, and he was very distressed that the operators had turned off the ECCS. Having designed the system, he understood the possible implications of turning it off. Dunn discussed this with Kelly, but it went no further at that time.

But on October 23, 1977, the same accident happened again at Davis-Besse. Once again, the operators became confused and turned off the ECCS system. Once again, the accident was caught in time to avoid a calamity. But this time, both Dunn at Babcock and Wilcox and Michelson in the ACRS took their concerns further.

Dunn went to Babcock and Wilcox's training department to ask what the operators were being taught. He was assured that the Davis-Besse actions were not in accordance with instructions. However, they had happened, and Dunn convinced Joseph Kelly to write an internal memorandum describing the accidents and pointing out the seriousness of the problem. Only one person responded to the memorandum, agreeing that operators were probably being mistrained.[37]

When several months went by without any additional response, Dunn tried writing a memorandum of his own, dated February 8, 1978, and addressed to James Taylor in the licensing unit. Part of it claimed:

> Had this event occurred in a reactor at full power with other than insignificant burnup [residual heat] it is possible, perhaps probable, that core uncovery and possible fuel damage would have resulted. . . . I believe this is a very serious matter and deserves our prompt attention and correction.[38]

That memorandum was never answered. Taylor told Dunn after the accident that he did not remember ever receiving it.

After the second accident, Michelson convinced Jesse Ebersole of the ACRS to write a letter to Sanford Israel at the NRC. Michelson and Ebersole decided that Israel was the person on the NRC staff most likely to appreciate the problem.[39] Israel was impressed, and he prepared a memorandum to be distributed within the NRC under the signature of Reactor Systems Branch Chief Thomas Novak.

However, the distribution of the memorandum stopped with the original list, and the problem did not even make the list of "generic problems" kept by the NRC. Since the problem described was not generic, it could supposedly be used in any licensing hearings for future Babcock and Wilcox reactors. However, almost no one knew about the Novak memorandum. Babcock and Wilcox was not told. Amazingly, even Jesse Ebersole in the ACRS was not told that his letter had generated an internal memorandum.

While Babcock and Wilcox did not know about the Novak memorandum in the NRC, they did finally receive a copy of Michelson's memo during 1978. It was routed to Bert Dunn, who was encouraged that someone else recognized and appreciated the problem that he had discussed. However, in the confusion of all the memoranda floating around, Dunn was not sure whether the problem had ever been communicated to the utility companies so that they could stop their operators from repeating the accident.

Over the next year, the issue stayed alive but partially dormant within Babcock and Wilcox. In February 1978 Dunn proposed a set of guidelines that could be passed along to the utility companies outlining under what conditions the ECCS should never be manually overridden. However, there was some disagreement within Bacock and Wilcox on whether Dunn's guidelines were too restrictive. So long as disagreement remained, Babcock and Wilcox did not pass the guidelines along to the operating companies. Unfortunately, that meant that the companies also had not heard that the pressurizer level could disagree with the pressure indicators.

On March 28, 1979, four and one-half minutes into a very confusing transient, operators once again faced a rising pressurizer level and falling pressure indicator. This was the fourth time that this condition had existed in five years. It was the third time in less than two years that it had happened at a Babcock and Wilcox plant. But this time, the plant was operating at virtually full power with old fuel.

The operators had never heard of Carl Michelson or the Novak memorandum. While some knew that there had been a transient at Davis–Besse, no one knew any details about it. They had no idea that people in the NRC and Babcock and Wilcox were aware that pressures and pressurizer levels could disagree. Instead, they understood that they had been taught the opposite. In addition, they had repeatedly been told that the pressurizer must never be run solid.

Despite five years of experience with this phenomenon and a pile of forgotten memoranda, the "regulators" or protectors of the technology had not managed to get the information where it was needed. Four and one-half minutes into the transient, the operators at Three Mile Island took the action that Dunn and Michelson had feared for almost two years. Their inaccurate training caused them to override a properly operating, properly designed safety system that would have prevented the accident.

Mistakes were unquestionably made during the Three Mile Island crisis for which there was no reasonable excuse. On the other side, some aspects of the Three Mile Island episode went better than we should have expected. On balance, it serves no real purpose to speculate about whether the people who participated in the Three Mile Island crisis were at fault. The reason why Three Mile Island happened is that somewhere, at some time, it was inevitable.

As stated earlier, we want to view nuclear energy and other forms of high technology as proactive. But high technology cannot operate that way. By its very nature, it must be reactive. High technology is a risk. In return for the creature comforts it offers us, it requires too much information in advance about what possible accidents can occur. It requires too delicate an interface between breakable machines and fallible people. It requires an accounting system on quality control and operational status that overloads us with information.

To be sure, there are lessons that can be learned from this accident that can improve the safety of nuclear energy. We can vent core vessels and redesign control rooms and retrain operators. Since the accident, the federal emergency management machinery has been entirely reorganized.

But there are other accidents at reactors still waiting to happen, and there is no guarantee that we will learn the lessons of Three Mile Island the first time we see them. As in all high technologies, one of the more effective techniques we can use to tell if safety systems have weaknesses is to wait for them to break.

At the beginning of this chapter, it was stated that there are two tenets to our faith in technology that are necessary to guarantee us progress with invulnerability. The first, that mechanical controls are available when the technology is truly dangerous, has not been addressed here. But the second, that there are people guaranteeing that the needed controls are in place, is an unreasonable expectation given the nature of high technology. It requires a knowledge of future technological possibilities that we cannot have without the experience of mechanical breakdowns.

A sense of invulnerability is still possible. As a society, we take automobile accidents and plane crashes and even small oil spills with some degree of fatalistic acceptance. We feel that we gain more from those technologies than we lose through occasional total collapses in the machinery.

But before Three Mile Island we were becoming increasingly lulled into the unrealistic assumption that in some technologies, every conceivable breakdown could be analyzed in advance and that some additional machine could be built to counteract it. Machines could then be built to counteract every conceivable breakdown of those monitoring machines. Then we assumed that we could understand all the possible ramifications, that at different times, different monitoring machines would be in operation. Then, without testing any of this through a failure of the entire system, we assumed that we could train people so well that they would know to intervene only when the unbreakable machine needed it and then do only what was required.

Any machine that we can build can break because we cannot know in advance every possible thing that can go wrong. But we as a society can still have our sense of invulnerability so long as we build only machines in which we can physically and emotionally tolerate occasional failures. If we as a society face this final issue and make a decision on whether we are willing to tolerate accidents in nuclear energy, then Three Mile Island has served us well.

NOTES

1. Edward Shils, *The Intellectuals and the Powers and Other Essays* (Chicago: University of Chicago Press, 1972), p. 230.

2. C.P. Snow, *Science and Government* (New York: New American Library, 1960), p. 9.

3. Donella H. Meadows, et al., *The Limits to Growth*, 2nd ed. (New York: New American Library, 1974); John Harte and Robert Socolow, eds., *The Patient Earth* (New York: Holt, Rinehart, and Winston, 1971); Erich Fromm, *The Art of Loving* (New York: Harper, 1956).

4. Richard Barnet, *The Economy of Death* (New York: Atheneum, 1970); John Kenneth Galbraith, *How to Control the Military* (Garden City, New York: Doubleday and Co., 1969).

5. Mary E. Ames, *Outcome Uncertain: Science and the Political Process* (Washington, D.C.: Communications Press, 1978), pp. 169–70.

6. Meadows, p. 29.

7. "Staff Report on the Generic Assessment of Feedwater Transients in Pres-

surized Water Reactors Designed by the Babcock and Wilcox Company," NUREG–0560 (Washington, D.C.: Nuclear Regulatory Commission, May 1979). "Investigation into the March 28, 1979 Three Mile Island Accident by Office of Inspection and Enforcement," NUREG–0600 (Washington, D.C.: Nuclear Regulatory Commission, August 1979).

8. Testimony of Harold Denton, President's Commission, August 23, 1979, pp. 36–37.

9. Transcript of President's Commission, August 23, 1979, pp. 36–126.

10. *Report of the President's Commission on the Accident at Three Mile Island* (Washington, D.C.: USGPO, October 1979), p. 24.

11. Ibid., pp. 61–67.

12. Paul Eddy, Elaine Potter, and Bruce Page, *Destination Disaster: From the Tri-Motor to the DC-10* (New York: Quadrangle Books, 1976), p. 180.

13. "Report of the Office of Chief Counsel on the Nuclear Regulatory Commission" to President's Commission, October 1979, pp. 54–55.

14. Ibid., p. 57.

15. Testimony of Jesse Ebersole, President's Commission, August 22, 1979, pp. 115–16.

16. Letter from Stephen H. Hanauer (ACRS) to Chairman Glenn Seaborg (NRC), on file at NRC reading room, July 17, 1969, p. 2.

17. "Report of the Office of Chief Counsel on the Nuclear Regulatory Commission," p. 62.

18. Ibid.

19. Ibid., p. 63.

20. Deposition of Robert Pollard, quoted in ibid., p. 66.

21. Deposition of Roger Mattson for the President's Commission, Bethesda, Maryland, August 6, 1979, p. 237.

22. Testimony of Victor Stello, President's Commission, August 23, 1979.

23. NRC internal memorandum from Steven Varga to "distribution," September 19, 1978.

24. "Technical Staff Analysis Report on Quality Assurance" to the President's Commission, October 1979, p. 17.

25. Ibid., p. 20.

26. Ibid., p. 19.

27. Ibid., pp. 80–84.

28. Ibid., p. 87.

29. Ibid., p. 55.

30. "Technical Staff Analysis Report on Selection, Training, Qualification, and Licensing of Three Mile Island Reactor Operating Personnel" to President's Commission, October 1979, pp. 81–86.

31. Ibid., p. 58.

32. Ibid., pp. 48–58.

33. Ibid., p. 49.

34. Scenario taken from testimony of Joseph LaFleur, President's Commission, August 22, 1979, pp. 5–11.

35. "Report of the Office of Chief Counsel on the Nuclear Regulatory Commission," pp. 72–73.

36. Ibid., p. 74.

37. Response quoted in "Report of the Office of Chief Counsel on the Role of the Managing Utility and its Suppliers" to President's Commission, October 1979, p. 154.

38. Reprinted in ibid, appendix L.

39. "Report of the Office of Chief Counsel on the Nuclear Regulatory Commission," pp. 74-75.

Chronology of the Crisis

WEDNESDAY, MARCH 28, 1979

4:00:36 A.M.	Polisher becomes isolated.
4:00:37 A.M.	Secondary feedwater pumps and turbines trip.
4:00:42 A.M.	Pressure relief valve opens.
4:00:44 A.M.	Reactor trips.
4:00:51 A.M.	Pressure relief valve fails to reclose on command.
4:01 A.M.	Pressurizer begins to fill with water.
4:03 A.M.	Operators override emergency core-cooling system.
4:05 A.M.	Pressurizer "goes solid," filling with water.
4:07 A.M.	Sump pump begins pumping water to auxiliary building.
4:08 A.M.	Closed auxiliary feedwater pump valves discovered and opened.
4:10 A.M.	Second sump pump engages.
4:15 A.M.	Quench tank rupture disk fails.
4:36 A.M.	Sump pumps shut off.
5:14 A.M.	B loop reactor coolant pumps shut off.
5:14 A.M.	Line printer jams, losing some information.
5:41 A.M.	A loop reactor coolant pumps shut off.
6:22 A.M.	Pressure relief block valve closes when open pressure relief valve discovered.
6:45 A.M.	Radiation readings soar in containment.
6:50 A.M.	Radiation alarms in auxiliary building.
6:55 A.M.	Site emergency declared and phone notifications begin.

7:02 A.M.	Pennsylvania Emergency Management Agency notified.
7:15 A.M.	Bureau of Radiation Protection notified.
7:24 A.M.	General emergency declared at plant.
7:40 A.M.	Bureau of Radiation Protection mistakenly told radiation in Goldsboro is at 10 rems per hour.
7:45 A.M.	Nuclear Regulatory Commission Region I learns of crisis.
8:00 A.M.	Thermocouple readings taken in cable spreader room, but not believed.
8:15 A.M.	Goldsboro readings discovered to be at background levels.
8:45 A.M.	NRC Region I team leaves for Three Mile Island.
8:50 A.M.	NRC's Bethesda emergency center operational.
10:00 A.M.	NRC reaches the plant. Department of Energy and Defense Civil Preparedness Agency on the way.
11:00 A.M.	Lieutenant Governor has press conference.
11:00 A.M. –1:30 P.M.	State monitors off-site releases.
11:30 A.M.	Depressurization starts at plant.
1:50 P.M.	Pressure spike in containment from hydrogen burn.
2:00 P.M.	Some movement seen in primary coolant loop.
2:30 P.M.	Lieutenant Governor meets with Met Ed officials.
4:30 P.M.	Lieutenant Governor has press conference to say Met Ed misrepresented the crisis.
5:00 P.M.	Repressurization started when depressurization failed.
5:08 P.M.	Secondary coolant loop restarted through A generator.
7:50 P.M.	Primary coolant loop restarted through A generator.

THURSDAY, MARCH 29, 1979

10:00 A.M.	William Dornsife goes to plant for Lieutenant Governor.
12:00 Noon	William Scranton tours plant.
2:45 P.M.	Waste water discharge started.
3:00 P.M.	Secretary of Health Gordon MacLeod receives call that he believes is recommending an evacuation.
5:00 P.M.	Sample of core-cooling water drawn in hot lab.
6:00 P.M.	NRC Commissioner Hendrie orders waste water discharge stopped.
6:00 P.M.	Core-cooling sample analyzed. Heavy core damage discovered.

8:00 P.M.	Battle between NRC and state begins over issue of restarting waste water discharge.
11:00 P.M.	Governor told of heavy core damage.
12:00 Midnight	Waste water discharge restarted.

FRIDAY, MARCH 30, 1979

Early morning	Makeup tank periodically vented. Control room and Bethesda discuss the possibility of depressurizing.
7:00 A.M.	James Floyd in control room begins continual venting of make-up tank.
8:00 A.M.	1200 millirem per hour puff recorded above vent stack.
8:34 A.M.	Floyd begins evacuation confusion by calling Dauphin County emergency management director. Calls continue to 9:00 A.M.
9:00 A.M.	Bethesda hears of 1200 millirem per hour release. Decides to recommend an evacuation.
9:15 A.M.	Harold Collins at Bethesda calls state with evacuation advisory.
9:20 A.M.	Radio advisory to citizens of Harrisburg on possible evacuation.
9:30 A.M.	Evacuation recommendation to Governor Thornburgh stopped by Bureau of Radiation Protection.
10:07 A.M.	Thornburgh talks to NRC Chairman Joseph Hendrie.
10:30 A.M.	President Carter calls Hendrie.
11:00 A.M.	Met Ed press conference plays down importance of bubble.
11:40 A.M.	Thornburgh and Hendrie agree to advisory for small children and pregnant women.
12:30 P.M.	Evacuation advisory released to public.
12:40 P.M.	Roger Mattson at Bethesda tells commissions of growing bubble problem for first time.
1:30 P.M.	At White House meeting, Thornburgh, Denton, and Jack Watson are designated as official government spokesmen.
2:15 P.M.	White House tells Thornburgh for first time that meltdown might occur.
2:30 P.M.	Denton arrives at Three Mile Island.
3:30 P.M.	NRC press conference at Bethesda alerts press to meltdown danger.

5:00 P.M.	Joseph Califano begins meeting of health officials.
5:15 P.M.	Jody Powell holds press conference at White House.
8:30 P.M.	Denton goes to Harrisburg to brief Thornburgh.
9:00 P.M.	Press conference held with Denton and Thornburgh.
9:30 P.M.	Hendrie asks Mattson to begin working on the possibility that an explosion might occur in the core vessel.

SATURDAY, MARCH 31, 1979

2:00 A.M.	Hendrie asks Vic Stello at Three Mile Island to begin working on the possibility that the bubble might explode.
3:00 A.M.	The Food and Drug Administration obtains approval for emergency production of potassium iodide.
4:00 A.M.	(Approximate) Bubble size estimated at 1000–1500 cubic feet and growing.
11:00 A.M.	Met Ed holds last news conference to say bubble is shrinking.
12:00 Noon	Califano holds "health cabinet" meeting. Decides to recommend evacuation to White House.
1:00 P.M.	Mattson in Bethesda gets first figures on oxygen in hydrogen bubble.
2:30 P.M.	Hendrie holds press conference. Says bubble could explode.
3:00 P.M.	Denton's advisors tell him bubble cannot explode.
5:30 P.M.	White House meeting where HEW spokesmen are told to go through NRC only.
9:00 P.M.	Associated Press story confirmed by Bethesda on bubble released to public.
9:30 P.M.	Denton tells Thornburgh twenty mile evacuation might be needed.
10:45 P.M.	Dauphin County threatens its own evacuation if it does not get help by 9:00 A.M.
11:00 P.M.	Denton and Thornburgh hold press conference to say explosion not possible and President Carter to visit next day.

SUNDAY, APRIL 1, 1979

Early hours	Bethesda experts decide bubble is already flammable.
1:30 A.M.	First potassium iodide arrives in Harrisburg.

10:00 A.M.	Scranton goes to Dauphin County emergency management center to stop evacuation threat.
9:00 A.M.	Hendrie and Mattson at Bethesda decide to go to Three Mile Island to warn of explosion.
1:00 P.M.	President arrives as NRC experts argue on lawn.
4:00 P.M.	Stello discovers mistake in explosion calculations.
9:40 P.M.	Thornburgh announces Governor's Action Center will take calls.
10:00 P.M.	Hendrie and Denton go to see Thornburgh. Bubble at 300 cubic feet.

MONDAY, APRIL 2, 1979

8:00 A.M.	Bubble down to 150 cubic feet. Most releases from bubble are steam.
8:30 A.M.	Met Ed employee leaks to press that bubble is gone.
11:15 A.M.	Denton holds press conference to say that bubble is small.
Afternoon	Food and Drug Administration passes recommendation to White House that potassium iodide be distributed immediately.
Evening	Denton and NRC agree to allow reactor to cool without depressurization.

Technical Glossary

Annunciator. Warning light and audible signal that a reactor system is shifting. Used to tell the operators of both system changes and malfunctions. TMI Unit II has 1600 annunciator lights and a signal that sounds much like a beeper.

Atmospheric dump valves. Steam from the secondary side of the steam generators can be routed through sixteen long pipes directly into the atmosphere rather than to the turbines when the normal line is blocked or suction is lost.

Auxiliary building. One of several buildings surrounding each containment building serving a number of purposes. At TMI, the auxiliary building holds the waste tanks and the vent header, among other systems.

Auxiliary feedwater pumps. Should flow stop through the main feedwater pumps in the secondary cooling system, these three pumps automatically engage to pump water from the condensate storage tank into the steam generator. Also called emergency feedwater pumps.

Boiling water reactor. A reactor in which the water cooling the core flows through the turbines and then back to the core. This design is produced in this country by General Electric and comprises about a third of U.S. reactors. See also *pressurized water reactor.*

Borated water storage tank. Tank containing about one-half million gallons of water with boron added, used to supplement or replace the flow of water through the primary coolant loop.

Boron. Chemical used in control rods and in suspension in the primary water. It has the characteristic that it absorbs neutrons freely and can regulate or stop a chain reaction.

Cable tray. A rack holding in place often hundreds of cables running from the control room to the components being controlled. Reactors have a history of fire in the cable trays in the spreader room, most notably at Brown's Ferry nuclear powerplant.

Chain reaction. A condition in the nuclear core where neutrons released through fission cause the fission of other nucleii so that the process is self-sustaining.

Code safety valve. Any of usually two or three valves atop the pressurizer that are designed as a safety feature to relieve serious overpressurizations quickly. Some reactors, including TMI, also have a pressure relief valve.

Cold leg. In the primary coolant loop, the cold leg is the exit pipe from the steam generator that then feeds into the reactor coolant pumps.

Commercial. An accounting shift in the status of a licensed reactor so that the cost of its construction can be included in applications for rate increases. This status is not directly tied to the operation of the plant or the generation of electricity.

Condensate polisher. See *polisher.*

Condensate pumps. Pumps in the secondary coolant loop that draw water from the condensor and rush it to the main feedwater pumps. The condensate booster pumps are also part of this section of the secondary coolant loop.

Condensor. In the secondary coolant loop, the condensor cools the steam from the turbines by transferring part of the heat to another water system. At TMI, this other water system is cooled by the large cooling towers.

Containment building. The building in pressurized water reactors where the primary coolant loop and the core are located. The building is constructed of concrete with heavy steel reinforcement so that explosions and radioactive releases inside should have minimal impact on the outside environment.

Control rods. Solid rods constructed of boron or sometimes of cadmium that are inserted into the core among the fuel assemblies to regulate or almost stop fission in the core.

Core. The accumulation of fuel assemblies where fission and heat emission take place, providing the source of heat to run the turbines.

Critical mass. The volume and concentration of nuclear fuel necessary to sustain a chain reaction. The precise quantity depends on the configuration and the purity of the fuel.

Emergency Core-Cooling System. A computer program that automatically overrides the control of pumps and valves in the reactor during serious losses of pressure to pump huge amounts of water to the core.

Fission. The splitting of the nucleus of an atom. Uranium 235 fissions to release two neutrons and a large amount of heat.

FSAR. See *Safety Analysis Report.*

Fuel. In American light water reactors, which constitute almost all commercial reactors in this country, the fuel used for fissioning is the isotope Uranium 235.

Fuel assembly. A grouping of about 150 to 200 fuel rods structured into an approximately square configuration when viewed from the end of the fuel rods.

Fuel-handling building. One of the supportive buildings at the TMI reactor where fuel is stored when not being used in the core.

Fuel rod. A hollow tube constructed of zirconium alloys that holds small pellets of the uranium fuel. Most fuel rods are about fifteen feet long. At TMI, they are between twelve and thirteen feet long.

Hot leg. The portion of the primary coolant loop pipe that feeds into the top of the steam generator. In Babcock and Wilcox reactors, this portion of the pipe is also sometimes called the candy cane because of its shape.

Integrated Control System. The computer program at a reactor that automatically runs the reactor during normal operation.

Isotopes. Variations in atoms caused by a difference in the number of neutrons in the nucleus. Most uranium is the isotope U–238, containing 92 protons and 146 neutrons. A more useful nuclear fuel for light water reactors is the isotope U–235, containing only 143 neutrons.

Let-down system. The group of valves and storage tanks that allows water to be removed from the primary coolant loop. This sytem is used more for adjustments in the loop than for relief during emergencies but is available for both purposes.

Light water reactor. A reactor that uses the regular isotope of water as the coolant and moderator.

Main feedwater pumps. The large pumps that return secondary coolant water from the condensate pumps to the steam generator, where it can absorb heat from the primary coolant loop.

Makeup system. The series of storage tanks and pumps that allow water to be added to the primary coolant loop. The major sources of this water are the let-down system or the borated water storage tank.

Megawatt. A measure of heat and electricity generation. Equal to one million watts. The TMI cores each produced almost 1000 megawatts of electricity.

Meltdown. An accident where a loss of coolant to the core results in the melting of the core's structure and finally of the fuel itself. It is a matter of speculation precisely what damage would occur in the complete meltdown of a commercial-sized reactor.

Moderator. The role of slowing the neutrons in the core to facilitate the fission of the fuel. In light water reactors, this role is fulfilled by the coolant water.

Neutron. An atomic particle normally contained in nucleus that has no electrical charge. Freed neutrons can stimulate fission in U–235.

Nuclear Regulatory Commission. The federal agency that licenses, regulates, and inspects commercial reactors in the United States. Before 1975, this agency was called the Atomic Energy Commission and included the role of fostering the development of commercial nuclear energy.

Nucleus. The center part of the atom, containing protons and neutrons.

Polisher. A unit that is designed to remove minerals and other contaminants from the cooling water of a reactor by having those impurities absorbed by a resin.

Power grid. The connection of electricity-generating sources and transmission lines that creates and delivers electrical current to the consuming public.

Polyurethane foam. The substance normally used as a stuffing to clog holes around electrical control cables. It is extremely flammable and has been ignited in reactors several times.

Pressure relief valve. The valve atop Babcock and Wilcox reactors that was used until the TMI accident to try to relieve enough pressure to avoid a reactor trip when the turbine tripped.

Pressurized water reactor. A light water reactor in which the water cooling the core transfers heat inside a steam generator to a secondary coolant loop. The secondary coolant loop drives the turbines for electricity generation.

Pressurizer. A large tank connected to one of the hot legs in a pressurized water reactor where a steam void is maintained during normal operation to allow flexibility in the volume of the primary coolant during temperature fluctuations.

Price-Anderson Act. A federal law first passed in 1957 limiting the liability from any one nuclear accident to $560 million, $500 million of which is covered by the federal government.

Primary loop. In a pressurized water reactor, the system in which water absorbs heat from the core, passes through the hot leg to a steam generator where it transfers heat, and then passes through the reactor coolant pumps back into the core vessel. The entire loop is within the containment building.

PSAR. See *Safety Analysis Report.*

PSI. Acronym for pounds per square inch, a measure of pressure. In this book, psi is always assumed to be pressure above atmospheric levels.

Quench tank. The tank at TMI that received the discharges of the primary coolant loop through the pressure relief valve and the code safety valves. This tank is more formally called the *reactor coolant* drain tank.

Reactor. A machine that uses the fission of a fuel to heat a coolant that then drives a turbine to create electricity.

Reactor coolant drain tank. See *quench tank.*

Residual heat removal pumps. Pumps normally designed to be used to supply coolant to the core when it is shut down and in the process of cooling. These pumps supply a larger volume of water than the makeup pumps, but cannot push water if the primary loop is at high pressure.

Resin. The chemical substance used to absorb impurities from the water as it passes through the polisher.

Safety Analysis Report. A report describing the safety features to be attached and the safety procedures to be used in a proposed reactor. The Preliminary Safety Analysis Report (PSAR) is filed with the NRC with the application for a construction permit. The Final Safety Analysis Report (FSAR) is filed with the NRC with the application for an operating license.

Scram. A reactor trip.

Secondary loop. In pressurized water reactors, the system in which water absorbs heat from the primary loop inside the steam generator, boils to steam, rushes through the turbines, converts back to water in the condensor, and returns to the steam generator.

Simulator. A mock control room in which a computer program responds to the actions of operators in the way that an actual reactor would presumably respond. The simulator can also be programmed to create malfunctions to which the operators must respond.

Spreader room. The room below a control room in which cables from the control panels are routed toward the components they control.

Steam generator. In pressurized water reactors, large tanks holding secondary water. Pipes of primary water lead through the generator and transfer heat to the secondary reservoir of water. The secondary water boils to steam and is released to flow through the turbines.

Sump. Water that collects in the basement of containment. It is drained to waste tanks in the auxiliary building by the sump pump.

Transient. A sudden shift in conditions within a reactor resulting from sudden changes in coolant flow. Transients occur in either the primary or secondary loops. They need not involve an actual trip of the reactor or turbine, but they usually involve the tripping of at least one of these components.

Trip. The shutting off of a unit, normally the turbine or the reactor, either intentionally or through a malfunction.

Turbine. Machines in which rushing steam turns blades and creates a rotating motion. In reactors, a string of high pressure and then lower pressure turbines are normally connected in series to generate more momentum.

Uranium. The fuel used in most reactors in the United States. See *isotope.*

Void. A collection of steam inside a cooling pipe or component. Under pressurized conditions, water has great difficulty flowing through a steam void.

Zirconium. The chemical element used as the base in coating fuel pellets and in constructing fuel rods and fuel assemblies. Zirconium is very effective in containing radioactive contaminants, but is relatively ineffective under conditions of great heat.

Acronyms

ACRS	Advisory Committee on Reactor Safeguards, attached to NRC.
AEC	Atomic Energy Commission, replaced by NRC and ERDA in 1975.
BRP	Pennsylvania's Bureau of Radiation Protection.
BWR	Boiling Water Reactor.
DCPA	Defense Civil Preparedness Agency.
DOE	Department of Energy.
ECCS	Emergency Core Cooling System.
EF-V	Emergency Feedwater-Valve.
EPA	Environmental Protection Agency.
ERDA	Energy Research and Development Agency, now consolidated into DOE.
FDA	Food and Drug Administration.
FDAA	Federal Disaster Assistance Administration.
FPA	Federal Preparedness Agency.
FSAR	Final Safety Analysis Report.
GPU	General Public Utilities.
HEW	U.S. Department of Health, Education and Welfare.
HUD	U.S. Department of Housing and Urban Development.
ICS	Integrated Control System.
IE	Division of Inspection and Enforcement in NRC.
IRAP	Interagency Radiological Assistance Plan.
LPZ	Low Population Zone.
NIOSH	National Institute of Occupational Safety and Health.
NRC	Nuclear Regulatory Commission.

NUREG   Staff report within NRC. Acronym unknown.
PEMA   Pennsylvania Emergency Management Agency.
PSAR   Preliminary Safety Analysis Report.
PSI   Pounds per square inch (always used in this work to reflect pressure above atmospheric levels).
PWR   Pressurized Water Reactor.
TMI   Three Mile Island.
TMI-II   Three Mile Island Unit II.
TVA   Tennessee Valley Authority.

Major Participants

Karl Abraham. Press relations agent assigned by NRC to Capitol in Harrisburg.

Robert Adamcik. Regional Director of the Federal Disaster Assistance Administration, who was eventually made the coordinator of the federal efforts in Pennsylvania.

John Aherne. NRC Commissioner.

Peter Bradford. NRC Commissioner.

Zbigniew Brzezinski. Advisor to President through National Security Council, which was President Carter's only contact with the crisis for the first two days.

Joseph Califano. Secretary of Health, Education and Welfare. Tried to intervene into recovery effort on behalf of health-related agencies.

Jimmy Carter. President of the United States. Began an active role on Friday, March 30. Visited the plant on Sunday, April 1.

Harold Collins. Member of NRC Emergency Management Team who made the evacuation recommendation to PEMA in Pennsylvania.

Douglas Costle. Administrator of the Environmental Protection Agency.

Walter Creitz. President of Metropolitan Edison Company.

Paul Critchlow. Press Secretary to Governor Thornburgh.

John Davis. Acting Director of NRC's Inspection and Enforcement Division. Also a member of NRC's Emergency Management Team.

Joe Deal. Head of the Department of Energy team that began off-site monitoring on the first day of the accident.

241

Harold Denton. Member of NRC's Emergency Management Team who went to the plant on Friday and became the official spokesman on the condition of the reactor.

Herman Dieckamp. President of General Public Utilities, the company that owns Metropolitan Edison.

William Dornsife. Member of Pennsylvania's Bureau of Radiation Protection. Pennsylvania state government's only nuclear engineer.

Richard Dubiel. Director of Chemistry and Health Physics for Three Mile Island. He took the radiation readings during the opening hours of the emergency.

Eugene Eidenburg. Deputy to Jack Watson. Helped coordinate the federal response from the White House.

Craig Faust. Control room operator on duty when transient began.

James Floyd. Supervisor of Operations for Unit II. He began the venting on Friday morning that led to the evacuation scare.

Edward Frederick. Control room operator on duty when transient began.

Stephen Gage. Assistant Administrator for EPA. Headed a monitoring team sent to Pennsylvania.

Charles Gallina. The health physics expert in the first NRC team to arrive at the plant.

Tom Gerusky. Director of the Bureau of Radiation Protection for the state of Pennsylvania.

Victor Gilinsky. NRC Commissioner. For the first two days, he kept Jessica Tuchman Mathews at the White House informed on the crisis.

Boyce Grier. Director of Region I for the NRC in King of Prussia, Pennsylvania.

Oran Henderson. Director of Pennsylvania Emergency Management Agency.

Joseph Hendrie. Charman and Commissioner of NRC.

John (Jack) Herbein. Vice-President of Metropolitan Edison who became the company's spokesman before the press.

James Higgins. Member of first NRC team to reach the plant.

Charles Kennedy. Director of Governor's Action Center, which became the rumor control center during the crisis.

Richard Kennedy. NRC Commissioner.

Mark Knouse. Administrative Assistant to the Lieutenant Governor. He was one of the first political figures informed of the crisis.

George Kunder. Shift Supervisor for Unit I, but quickly went to Unit II to help.

Gordon MacLeod. Secretary of Health for Pennsylvania. Received a call from NIOSH on Thursday that he believed recommended an evacuation.

Jessica Tuchman Mathews. Staff member of National Security Council, and the White House source of information on the crisis for the first two days.

Roger Mattson. Engineer for NRC's Emergency Management Team who decided that hydrogen bubble was explosive.

Carlyle (Carl) Michelson. Engineer for TVA who predicted a TMI-type accident based on an earlier accident at the Davis-Besse nuclear reactor.

Gary Miller. Station Manager for Three Mile Island, and in charge of the recovery effort for most of the first day.

Kevin Molloy. Emergency Management Director for Dauphin County, which contains both the reactor and the city of Harrisburg. Gave radio advisory to the city on Friday morning.

Norman Moseley. Member of NRC's Emergency Management Team.

Jody Powell. Press Secretary to President, and official spokesman for federal recovery effort after evacuation scare.

Margaret (Maggie) Reilly. Member of Pennsylvania's Bureau of Radiation Protection, who was instrumental in monitoring the crisis and interpreting it for political decision-makers.

Michael Ross. Operations Supervisor for Unit I. Gary Miller put him in charge of the Unit II control room for much of the first day of the crisis.

Fred Shiemann. Shift Foreman for Unit II at time transient began.

William Scranton. Lieutenant Governor of Pennsylvania, and principal political decision-maker involved until Thursday evening. Also chaired State Emergency Council and State Energy Council.

Victor Stello. NRC engineer working under Denton at the plant who discovered the error in the hydrogen explosion calculations.

William Thornburgh. Governor of Pennsylvania.

John Villforth. Director of Food and Drug Administration. Instrumental in arranging for production of potassium iodide.

Jack Watson. Presidential advisor who headed White House response. Also, he is head of the White House Interagency Task Force on Emergency Planning.

William Zewe. Shift Supervisor in office behind Unit II control room when transient began.

Name Index

Subject Index

About the Author

Daniel Martin has a Ph.D. in Political Science from the Maxwell School at Syracuse University, where he specialized in government regulation of science and technology. He is currently teaching public administration at the University of Baltimore. Among his previous research interests were the government's efforts to build an American supersonic transport, and cargo door problems in the DC-10 aircraft. Professor Martin has also studied emergency planning around nuclear powerplants for the American Society for Public Administration and the Council of State Governments.